China Insights

This book series collects and presents cutting-edge studies on various issues that have emerged during the process of China's social and economic transformation, and promotes a comprehensive understanding of the economic, political, cultural and religious aspects of contemporary China. It brings together academic endeavors by contemporary Chinese researchers in various social science and related fields that record, interpret and analyze social phenomena that are unique to Chinese society, its reforms and rapid transition. This series offers a key English-language resource for researchers and students in China studies and related subjects, as well as for general interest readers looking to better grasp today's China.

The book series is a cooperation project between Springer and China Social Sciences Press of China.

More information about this series at http://www.springer.com/series/13591

Jiahua Pan

China's Environmental Governing and Ecological Civilization

中国社会科学出版社
CHINA SOCIAL SCIENCES PRESS

Jiahua Pan
Sustainable Development-General Research
Chinese Academy of Social Sciences
Beijing, China

This book is published with financial support from Innovation Project of CASS.

ISSN 2363-7579 ISSN 2363-7587 (electronic)
China Insights
ISBN 978-3-662-47428-0 ISBN 978-3-662-47429-7 (eBook)
DOI 10.1007/978-3-662-47429-7

Library of Congress Control Number: 2015951962

Springer Heidelberg New York Dordrecht London

Printed on acid-free paper

Springer-Verlag GmbH Berlin Heidelberg is part of Springer Science+Business Media (www.springer.com)

Preface

Modern China, since the Opium War, has dropped far behind and suffered from attacks and bullies. "Inferior technologies, systems and culture" becomes a prevailing cultural psychology among most Chinese. Changing the pattern that "China is comparatively weaker than western countries" and reinvigorating China start from cultural criticism and innovation. Since then Chinese has begun to "open their eyes to see the whole world around" and learnt from Japan, Europe, America and Soviet Russia. We are in tension and anxiety for long, dying to get over from being behind and bullied, poor and weak and surpass Western powers. As it were, in the pursuit of dream of power nation and national rejuvenation for the last century, we focus on understanding and learning but do a little even nothing to get understood. It has no significant changes even in the modernization evolvement after China's reform and opening up in 1978. Such phenomenon is well demonstrated by a great amount of translation of western writings in 1980s and 1990s. This is how Chinese perceives the relation between China and the world in modern times.

At the same time when Chinese is pursuing for the dream of power nation and national rejuvenation in modern times, they are also trying to seek for **DAO** to rescue from national subjugation and get wealthy and powerful by "material (technology) criticism", "institutional criticism" and "cultural criticism". **DAO** is firstly a philosophy, of course, as well as a flag and a soul. What philosophy, what flag and what soul are critical to resist national destruction and get wealthy and powerful? Over a century, Chinese people have kept exploring and attempting in disgrace, failure and anxiety. They have undergone the phase of "western learning for practical application while Chinese learning as basis", the failure of constitutional monarchy practice, the bankruptcy of western capitalist politics, and the major frustrations of world socialist movement in the early 1990s. But finally they have harvested Chinese revolution success, national independence, and liberation, especially combined scientific socialist theoretical logic with Chinese social development history logic and found out a Chinese socialism modernization path—a path of socialism with Chinese characteristics. After the reform and opening up in recent 30 years, China's socialism market economy has got rapid development; economic, political, cultural and social construction have made tremendous achievements; comprehensive national strength, cultural soft strength and international influence have substantially improved; the socialism with Chinese characteristics has made huge success; although not perfect, the institutional systems are generally established. China, a dream catcher for last century, is rising among the world nationalities with more confidence in unswerving path, theory and institution.

Meanwhile, we should be aware that the long-developed perception and the cultural mental habit of learning from the West constrain Chinese from showing "historical China" and "contemporary real China" to the world, although China has risen to be one of the current world powers. Western people and nationalities, influenced by the habitual history pattern that "China is comparatively weaker than western countries", rooted in Chinese and western cultural exchange, have little knowledge of China's history and contemporary development, let alone China's development path, as well as cognition and understanding of philosophical issues such as scientificity and effectiveness of "China theory" and "China institution" on China, and their unique values and contributions to human civilization. The omission of "self-recognition display" leaves much room for malicious politicians to wide spread "China collapse", "China threat" and "China national capitalism".

During the development process of "crossing a river by feeling the way over the stones", we concentrate more energy on learning from the West and understanding the world, and get used to recognize ourselves by western experience and words, but ignore "self-recognition" and "being understood". We blend ourselves into the world more forgiving and friendly, but are not objectively or actually understood. Hence, just and responsible academic and cultural researchers should undertake the important assumptions to summarize DAO of successful socialism with Chinese characteristics, narrate China stories, illustrate China experience, use international expressions, tell the world the real China, and let the world acknowledge that western modern pattern is not the end of human history and the socialism with Chinese characteristics is also a valuable treasure of human thoughts.

Chinese Academy of Social Sciences organizes leading experts and scholars and some outside experts to write series of *Understanding China*. These books introduce and summarize China path, China theories and China institutions, and contain objective description and interpretation to modern development in aspects of political system, human right, law governance, economic system, economics, finance, social administration, social security, population policy, value, religious faith, nationality policy, rural issue, urbanization, industrialization, ecology and ancient civilization, literature and art.

These published books are expected to let Chinese readers have better understanding of China's modernization process in the last century, and more rational concepts on current troubles; intensify the overall reform and national confidence; agglomerate consensus and strength for reform and development; improve foreign readers' understanding of China and create more favorable international environment for China's development.

Beijing, China
January 9, 2014

Contents

Foreword

The enormous gap between China's fragile ecosystems and the beautiful Chinese Dream means the long way ahead, the direction of efforts, and the requests for hard work. China's urbanization and industry-dominated development is rapid, at large scale and lasts a long period. As "the world's factory," China's huge resource supply, processing and manufacturing, energy consumption, product consumption, and pollutant emissions and discharges are exceeding the limits of its environmental capacity. The interconnected security of food, water, ecosystem, energy, climate, and environment requires China to speed up its transition toward a green economy and society. The international community also expects China to play an important role in global climate governance, energy reform, and development path change. China's comprehensive, large-scale, and innovative efforts for building a society of ecology civilization creates a brand new development paradigm for the harmony between human society and nature and the sustainable development of human society and economy.

The contents of constructing an ecological civilization mainly include ecosystem protection, pollution control, improving the efficiency of natural resource utilization, and integrating them in economic, political, social, and cultural activities. Obviously, ecological civilization is not limited to ecosystem protection. Instead, it has more comprehensive and fundamental social and economic implications. From the historic perspective of social development, ecological civilization with its long history is of great significance in modern society and represents the direction of future development. Ecological civilization is not empty propaganda. Instead, it contains measurable indicators and evaluation systems. During its efforts for ecological civilization building, China has created a complete set of governmental institutions and policies and made rapid progress and remarkable achievements in energy efficiency improvement, greenhouse gas emission reduction, pollution control, ecological rehabilitation, renewable energy utilization, and shifts toward green consumption. In face of the various serious challenges of industrialization and urbanization, China urgently needs to upgrade and transform its industrial civilization, to change from conquering and changing nature to respecting and protecting nature, from focusing on profit maximization to human-oriented sustainable development. It has to fundamentally transform its production patterns and life style, realize the harmony between human society and nature, to accelerate the construction of ecological civilization, and to lead the global transition toward ecological civilization.

Harmony between human and nature is the essence of ancient Chinese philosophy on ecological civilization. Human beings originated from nature, therefore maintaining harmony between human and nature is a basic principle. It also provides the solution in times of conflicts between human and nature. It is necessary to maintain harmony between human and nature in order to avoid and solve various problems and conflicts and achieve sustainable development. Therefore, harmony between human and nature represents the target and future development direction. To respect, accommodate, and utilize nature, human beings need to understand nature. Social economic development is human activities. Industrialization and urbanization are necessary to satisfy human demand for commodities. The process of ecological civilization building is also the process of understanding and putting into practice the concept of harmony between human and nature. China's efforts to build an ecological civilization are determined by the country's special natural environment. West China is characterized by fragile ecosystems, and it is also the natural defense for ecosystems in East China. East China lacks mineral resources, but its population and economic activity density is much higher. The Hu Line between West China and East China[1] in fact indicates the different climate capacities of the two regions due to water availability. Therefore, the realization of the beautiful Chinese Dream is subject to the hard constraints of natural environmental capacity.

From the perspective of human development needs, China's modernization process started later than many other countries; the country has a big population; and different parts of the country are at different development stages and of different development speeds. China's industrialization has experienced rapid progress over three decades since the late 1970s and overall China has entered the late stage of industrialization. Some of its advanced regions have even entered the phase of post-industrialization. However, China's industrialization has caused serious resource shortage, ecosystem damage, and environmental pollution, and it is impossible for China to continue the traditional industrialization based on intensive natural resource inputs and low technical content. The current status of China's industrial development and the natural resource and environment constraints require rapid transformation of Chinese industrial activities. China's urbanization rate is about at the global average; however, the quality of China's urbanization is relatively low. China faces the huge challenges of improving the quality and level of its urbanization under the existing tight constraints of national natural resource shortage, but new urbanization also provides important opportunities. It offers an engine for sustained and strong economic growth, effective approaches for improving ecological

[1]The Hu Line, also known as the Heihe-Tengchong Line or Aihui-Tengchong Line, was a straight line proposed in 1935 by Huanyong Hu, a Chinese demographer and geographer. The Line divided China into two parts of different population density.

efficiency, and a platform for the improvement of living standard. At the age of economic globalization, China has the special position of "the world's factory," a competitive big country specializing in manufacturing products with low and medium technology contents. As "the world's factory," China needs to utilize both national and international resources and markets to improve its general welfare and its ecosystems. The economic and asset value of natural resources depends on the coexistence of multiple resources. It is hard to assess the natural value or socioeconomic productivity of a single natural resource, for example, water or land, in absence of the natural and social context. In a desert, land is of low natural productivity; as a consequence, its socioeconomic value is low. However, to some extent, energy service can improve the efficiency of natural and socioeconomic production. As a result, the security of water, land, food, and energy is closely interconnected. Water security inevitably affects food security; energy security is closely related to water security. In pursuit of ecological security, China needs to take into account the interconnections among its natural resources. China is the biggest emitter of greenhouse gases (GHG) and its per capita GHG emissions are about at the level of the EU and its per capita GDP is almost at the level of mid-income countries. Therefore the international community anticipates China to reduce its GHG emissions. China needs to and can make active emission reduction contributions, but such contributions do not necessarily mean that China has to dramatically reduce its aggregate GHG emissions in the short term. It is more significant that China contributes to climate change mitigation in the areas of renewable energy use, land use, and forestry as well as low-carbon buildings. Industrial revolutions are the heralds of industrialization progress. Steam engine and information technology are the key technology breakthroughs of the first and the second industrial revolution. What is the key technology breakthrough of the third industrial revolution? Some people say it is three-dimensional (3D) printing. But 3D printing is only the combination of mechanic production and information technology, it does not represent a revolutionary technological breakthrough. Renewable energy production and supply can fundamentally lead to energy sustainability and represents a revolutionary technology breakthrough. Moreover, unlike the previous industrial revolutions which were based on a single technology innovation, renewable energy revolution is a comprehensive and large-scale technology revolution combining multiple technologies and covering a variety of energy sources. China has already started and is continually pushing forward the renewable energy revolution.

Economic growth is an important measurement and often the target for social development. However, advanced economies have already reached development saturation and the internal dynamics and the space for their economic extension and growth inevitably become weaker and smaller. As mature and saturated economies, Europe and Japan find the space for their economies' extensive growth limited or even diminishing. This is because in rich societies, the material needs of human beings have been satisfied, further growth is slow, even unnecessary or with negative impacts. This means that even in absence of resource and environment constraints, an economy does not necessarily maintain rapid growth for an unlimited period. Slow economic growth or even stagnation in developed countries is a natural tendency. Under resource and environmental constraints, a "steady-state economy" without size expansion can help realize the harmonious coexistence between human and nature. The slowing down of economic growth in China is unavoidable. Moreover, China needs to be ready to enter the era of "steady-state economy." Urbanization, industrialization, resource and environment constraints for economic growth and sustainable energy services are impossible to occur in the age of agricultural civilization. Human society needs a new form of social civilization to upgrade and transform industrial civilization. The philosophy and ethics behind ecological civilization is not utilitarianism, the philosophy and ethics of industrial civilization. It is respecting nature and human and pursuing ecological justice and social justice. Ecological civilization emphasizes not only the economic efficiency of profit maximization, but also the ecological efficiency of natural harmony and the social efficiency of social harmony. It also needs technology innovation and encourages technology innovation, but the technologies are not simply for the purposes of profit maximization and economic efficiency. More importantly, they are for the purpose of supporting the realization of quality and healthy human life and ecological environmental sustainability. GDP is the core measurement for social progress and economic development in industrial civilization. What is the core measurement for societies of ecological civilization? Quality life and environment, health, green and low-carbon development are the major elements of ecological civilization building and the key values of Chinese socialism. Industrial civilization is based on market and legal mechanisms. Ecological civilization societies do not disregard the legal and market mechanisms of industrial civilization societies. Instead, they add ecological civilization contents to market and legal mechanisms. For example, ecological boundaries are defined, ecological compensation is carried out, and natural resources are included in national balance sheet. Through its ecological civilization building practices, China has accumulated many successful experiences. These experiences can directly contribute to global ecological security; more importantly, they are pilot activities for the global transition toward ecological civilization. China is entering the new era of building an ecological civilization.

Ecological civilization as a new paradigm for green economy requests experimentation in practice and academic research. It is especially important to make theory innovations, identify methodologies, conduct detailed case studies, and perform data analysis. The topics requesting extensive theory studies include in-depth and systematic discussions on the concept of ecological civilization, refining the scientific connotations of ecological civilization, especially clarifying the connections and differences between ecological civilization and industrial civilization. It is also necessary to highlight the advantages of ecological civilization, to reveal the inevitability of China's pursuit for ecological civilization and the universal applicability of ecological civilization, and to explain its significance as a new phase in human socioeconomic development. In methodology research, it needs to be recognized that the GDP-focused system of national account reflects the utilitarianism ethical foundations and profit maximization target of industrial civilization. Ecological civilization societies need a new scientific and objective indicator and evaluation system. Natural resource asset evaluation and accounting is important.

However, the monetary values of natural resources are subject to the influences of market prices; therefore, the market value of natural resource assets may not reflect changes in natural resource sustainability.

In the scientific study about development and resource constraints, it is necessary to examine environmental status and carrying capacity, industrialization, urbanization, interconnections among natural resources, renewable energy revolution as well as the "ceiling" effects of development from the perspective of harmony between human and nature, so as to reveal the science and objectivity of the ecological civilization paradigm through statistical analysis and case studies. On the basis of theory study, methodology system establishment as well as analysis on the major challenges in reality, it can be concluded that China's transition toward green growth indicates it is entering the new development phase of ecological civilization.

List of Figures

List of Tables

Chapter 1
Ecological Capacity Profile and Adaptation

Ecological capacity, as a concept of volume, can be understood either in absolute terms or in relative terms. The green transition of China, to a large extent, is not an active choice but a reactive response. The natural resource endowments of China are specific, and their spatial distributions are predetermined. Obviously, climate conditions directly influence the productivity of natural systems. Therefore, the so-called environmental carrying capacity is in fact a kind of climate capacity. The geographic distribution of Chinese population and economy is subject to the constraints of climate capacity geographic distribution. The Chinese proverbs "people are shaped by the natural environment around them" and "timely wind and rain bring good harvest" are descriptions about climate capacity and their great importance. The large-scale human migrations in Chinese history and the current "ecological migrants," in many cases, are climate migrants because human activities exceeded the climate capacity or environmental carrying capacity of specific regions. Environmental carrying capacity is the basis and constraint condition for the transition toward ecological civilization. Respecting and accommodating nature mean the transition from the passive adaptation to the impacts of environmental constraints to the active adaptation of keeping human activities within environmental carrying capacity.

1.1 The Spatial Distribution of Resources

Carrying capacity can be defined from different aspects, including resource carrying capacity, ecological capacity, and environmental carrying capacity. It is the accommodation capacity or supporting capability and rigid. The absolute resource shortage in the Malthusian Theory of Population and the "Limits to Growth" theory by the Club of Rome regarded environmental carrying capacity as an absolute and impassable constraint on population size and consumption level. The Chinese government has set a threshold red line of maintaining a minimum of 1.8 billion

J. Pan, *China's Environmental Governing and Ecological Civilization*,
China Insights, DOI 10.1007/978-3-662-47429-7_1

hectares of arable land in the country. This is also a threshold for guaranteeing sufficient national capacity for grain production.

In a given area, climate and geographic conditions, as exogenous variables, are generally stable; therefore, normally the area's ecological carrying capacity and population supporting capacity are also constant. Once socioeconomic demand, because of population increase, exceeds the carrying capacity of natural ecosystems, the ecosystems and the socioeconomic systems they support will collapse. The theory of carrying capacity emphasizes that human activities must not exceed the absolute upper limit of specific ecosystems. Essentially, this means the existence of a long-term and reasonable upper limit to the expansion for human sustainable development. Traditionally, studies on carrying capacity are mainly from two perspectives. One is the ecological perspective: the studies assessed such concepts as "ecological carrying capacity," "environmental capacity," and "ecological footprint" from the constraints of natural resources and physical environment. As the Earth is the only home for mankind, despite technology progress, social system changes, and alterations in people's consumption and living styles, there are limits to the availability of some nonrenewable and irreplaceable resources that are indispensable for human survival. Although various studies are common in stressing the limits of natural carrying capacity, there are some differences among them due to their focuses on different key constraints. For example, the theory about "ecological carrying capacity" from the perspective of ecological balance calculates the supporting capacity of ecosystems worldwide in providing materials and ecological services for human activities. The theory about "environmental capacity" estimates the limits for human activities from the angle of environmental quality, especially air and water quality. It considers environmental capacity as the maximum quantity of specific pollutant a specific natural environment can absorb without causing damage to human survival and natural systems.

Meanwhile, carrying capacity can also be relative and depends on such factors as technologies, social choices, and values and has relative limits and ethical features. The theory of differential rent[1] formulated by the English classical economist David Ricardo is a typical theory about relative shortage of resources and highlights the flexibility of environmental capacity. Ricardo believed that the amount of fertile land is limited; however, the amount of poor-quality land is unlimited. Given capital inputs and technology progress, marginal or poor-quality land would constantly enter the market and satisfy demand.

The United Nations World Commission on Environment and Development from the perspective of technology and development defined "population supporting capacity" of the environment as the limitations imposed by the present state of technology and social organization on the ability of environmental resources and

[1] The differential rent theory states that the rent of a land site is equal to the economic advantage obtained using the site in its most productive use, relative to the advantage obtained by using marginal (i.e., the best rent-free) land for the same purpose, given the same inputs of labor and capital. The term "differential rent" was formulated by David Ricardo around 1809 and presented in its most developed form in his magnum opus, *On the Principles of Political Economy and Taxation*.

the biosphere to "meet the needs of the present without compromising the ability of future generations to meet their own needs." Sustainable development economics focuses on the question of global ecosystems' population supporting capacity or the thresholds for global population supporting capacity and development. Kenneth E. Boulding, an American economist, developed the term "spaceship economy." According to him, due to boundary limitations of the Earth, economic activities, in principle, cannot and should not be an open and illimitable "cowboy economy." The resources human beings can use are limited, so is the space into which human can dump wastes. Therefore, the Earth is a single spaceship, and the world economy is a closed "spaceship economy."[2] Population and economy growth, if out of control, will sooner or later exhaust the limited resources on the spaceship, the Earth. Therefore, the "population supporting capacity" of an ecosystem not only depends on the quantity and quality of natural resources and environment but also is subject to the influences of human activities and the development path choice, production patterns, and consumption styles of human societies. Due to the focuses of different disciplines and the application of different analysis approaches, many studies neglect the complicated interrelations between human activities and the population supporting capacity of ecosystems and environment. Particularly, under the background of climate and environmental change, rapid socioeconomic development (especially quick urbanization) further increases the uncertainties and complexities of the human–nature interrelations and makes it more difficult to assess the population carrying capacity of specific ecosystems and the environment.

The geographic distribution of different natural environment and resources predetermines the major differences in the population supporting capacity of different areas on the Earth. Rain forest, meadows, and deserts have far different population supporting capacity. China has a vast territory, and the natural productivity of resources and environment varies widely among different parts of China. In 1935, Huanyong Hu, a Chinese scholar on population and geography, found a straight population density line (around 45°, known as the Hu Line) starting from Aihui in Heilongjiang Province to Tengchong in Yunnan Province.[3] The southeast of the line covered 36 % of China's land area but homed 96 % of the Chinese population, while the part to the northwest of the line covered 64 % of the Chinese land area. However, only 4 % of the national population lived there. The difference between the average population densities of the two regions was as big as 42.6:1.

Even after 80 years, the population distribution divided by the Hu Line still persists. On the population map developed by Huanyong Hu in 1935 (see Fig. 1.1a), 96 % of the population lived in the southeast of the line; the regions with the highest population density were coastal areas in Southeast China; especially, the Yangtze Delta was the biggest region with high population density. Such a distribution of Chinese population existed in the time of agriculture economy and low productivity,

[2] Boulding, K. E. (1966). The economics of the coming spaceship Earth. In H. Jarrett (ed.), *Environmental Quality in a Growing Economy* (pp. 3–14). Baltimore: published for Resources for the Future, Inc. by The Johns Hopkins Press.

[3] Huanyong Hu. (1935). Population distribution in China. *Journal of Geographical Sciences*, 2.

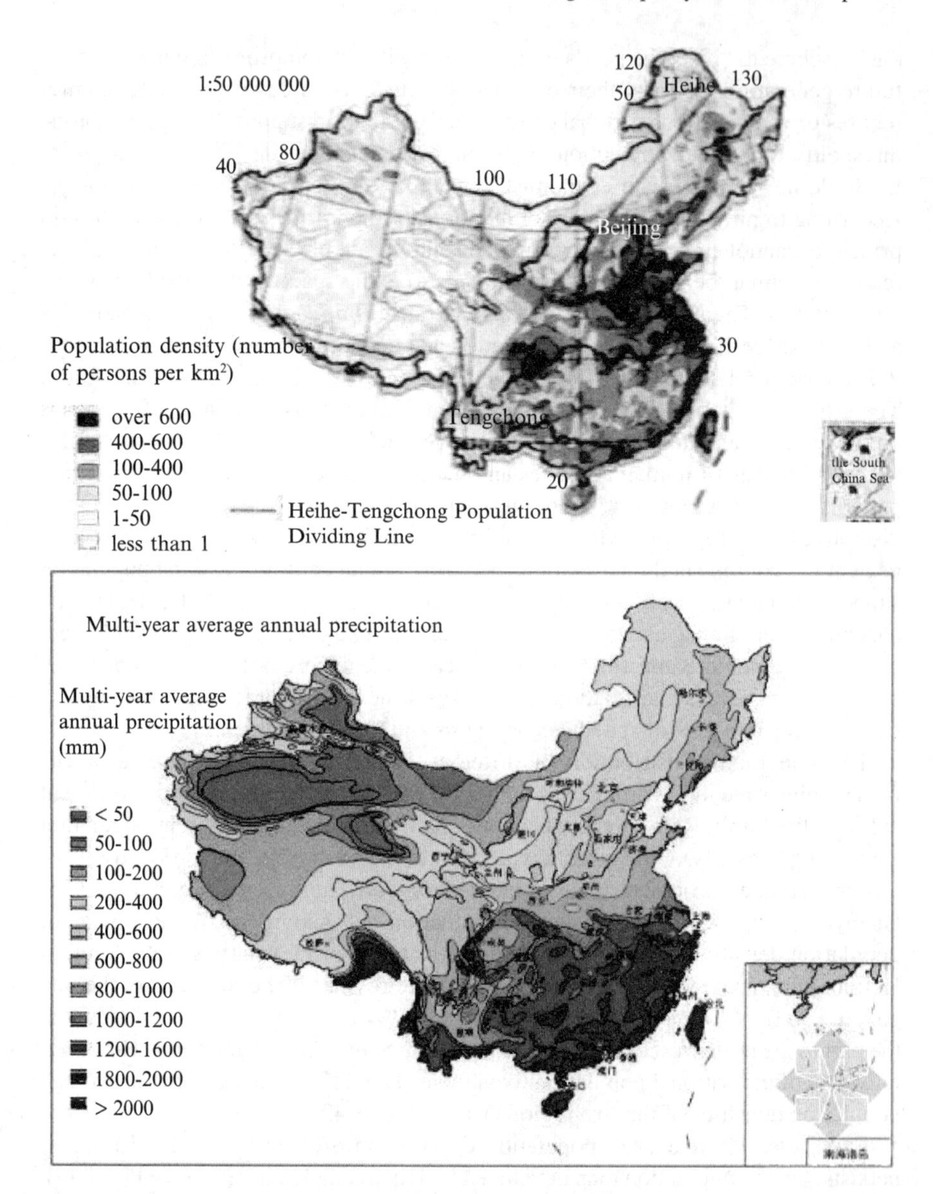

Fig. 1.1 Geographic distribution of population and precipitation in China. (**a**) Population distribution in China and (**b**) distribution of annual precipitation

and it remains almost the same in the current industrial society of much higher socioeconomic development level and much more advanced technologies. Results of the third and the fourth national censuses happened in 1982 and 1990 indicated no major change in the geographic distribution of Chinese population. For example, in 1982, Southeast China accounted for 42.9 % of the Chinese land area and occupied by 94.4 % of the Chinese people, and in 1990, its share in national population was 94.2 %. This means that the big differences in population shares did not change much during the 55 years before 1990. The fifth China census taking place in 2000 found that the population shares on the two sides of the Hu Line were, respectively, 94.2 % and 5.8 %, indicating almost no change in the 1990s. Although the population shares of the two regions barely changed over time, the number of Chinese people had more than tripled since the discovery of the Hu Line.

The Hu Line is both a line of population density and a natural geographic boundary that almost overlaps the 400 mm annual precipitation line (see Fig. 1.1b) which is also the dividing line between semi-humid southeast and semiarid northwest in China. Regions on the northwest of the Hu Line have an annual precipitation less than 400 mm and are plagued by desertification. The topographies and natural conditions are unfavorable for vegetation growth; therefore, the productivity of local natural ecosystems and ecological capacity are low, and the region is unable to support a large number of population who need abundant and reliable food supply. The region is dominated by grassland, deserts, plateaus, and high mountains; the primary economic sector is nomadic animal herding, and the intensity of local socioeconomic activities is low. In the southeast of the Hu Line, the topography mainly consists of plains, variegated rivers, hills, karst, and Danxia landforms. The region as a whole has abundant precipitation, a topography favorable to vegetation growth, high biodiversity, highly productive ecosystems, and natural environment. Therefore, it is suitable for farming activities and has sustained an advanced agricultural civilization for many centuries.

The main features of resources and environmental distribution in different parts of China, apart from the overall features reflected by the precipitation-based Hu Line, also include topography and vegetation conditions which are important factors affecting human settlement. Feng et al. divided China into 1 km × 1 km squares[4] based on the geographic information system and technology and quantitatively assessed different regions' suitability for human settlement to reveal the natural setup and regional variations of human settlement environment in China. Their study results showed that overall the environmental index for human settlement is the highest in the coastal areas in Southeast China and declines gradually toward the inland regions in Northwest China. A region's population density is closely related to its human settlement environmental index level and reflects the suitability of local natural environment for human life. The size of regions suitable for human settlement is 430.47 × 104 km^2 in China, accounting for approximately 45 % of the country's land area and 96.56 % of the Chinese people living in these regions; among them, more than 3/4 of the Chinese population concentrate on about 1/4 of the country's land which is highly or moderately suitable for human settlement. The marginal regions for human settlement cover 225.11 × 104 km^2 of land and are home to 41.12 million people or 3.24 % of Chinese population. With an average population density

[4] Zhiming Feng, Yan Tang, Yanzhao Yang, & Dan Zhang. (2008). The establishment and application of GIS-based Human Settlement Environmental Index in China. *Journal of Geographical Sciences*, (12).

of 18 people per square kilometer, they are the transitional areas between regions suitable for human settlement and those unsuitable for human settlement. The rest 304.42×104 km^2 or 31.71 % of the country's total land area are unsuitable for human settlement. Only 2.49 million people, less than 0.2 % of the Chinese people, live in these regions. The average population density in these regions is less than one person per square kilometer, and vast areas are totally unsuitable for human settlement, and the majority of these regions are no-human zones.

The spatial distribution of ecological resources also determines the location and distribution of Chinese cities and industrial activities. The Yangtze Delta, the Pearl River Delta, and the Bohai Sea rim area, which are homes to urban agglomerations each with over 50 million people, are all located in the coastal areas of East China. The locations of urban clusters each with over 10 million people, including the middle and lower reaches of the Yangtze River, Chengdu–Chongqing, and Harbin–Changchun, are also in thc southeast of the Hu Line and near coastal areas. The biggest provincial capital cities in Northwest China, Lanzhou, Urumqi, and Hohhot, only have around three million people each. In terms of industrial activity distribution, Northwest China is the production center of energy and raw materials. If they were suitable for human settlement, the huge energy projects of West–East Electricity Transmission and West–East Gas Pipeline wouldn't be necessary. Agriculture, which highly depends on climate and other natural productivity conditions, bears even more distinct features of "East" and "West" China. Based on the combinations of temperature and precipitation levels and variations, China can be generally divided into two parts along the Hu Line: the eastern monsoon zone is dominated by farming activities; in the western region which includes northwest arid zones and the Qinghai–Tibet alpine frigid zones, the main agricultural activity is nomadic animal herding. This shapes the prominent feature of "farming in the east and herding in the west" in agriculture activity distribution in China.[5] In July 2005, the Chinese State Forestry Administration put forward the forestry development strategy for ecological civilization building which is "expansion in the east, improvement in the west, utilization in the south, and rehabilitation in the north." The objectives of the strategy are to increase the national forest coverage rate to 23 % by 2020 and to realize dramatic improvement of ecosystems nationwide. Expansion in the east means to expand the size and increase the density of forest in coastal areas with advanced economy in Southeast China. Improvement in the west means to improve the fragile ecosystems in Western China. Utilization in the south refers to improving the quality and economic returns in the warm areas with sufficient sunlight and precipitation in South China. Rehabilitation in the north refers to protecting and rehabilitating the primitive forests in Northeast China and Inner Mongolia.[6] Obviously, the different strategies for different regions are also based on the regional differences highlighted by the Hu Line.

[5] BaojianQiu. (1986). A proposal for comprehensive planning on natural functions of agricultural land in China. *Journal of Henan University* (natural science version), (1).

[6] Such regional differentiation was key elements of the 11th Five-Year Plan (2006–2010) made by the State Forestry Administration. State Forestry Administration, 2006.

The National Plan for Main Functions of Different Zones[7] include the ecological security program of establishing "two defenses and three belts," the Qinghai–Tibet Plateau Ecological Defense, the Loess Plateau–Sichuan–Yunnan Ecological Defense, the Northeastern Forest Belt, the Northern Desertification Control Belt, and the Belt of Hills and Mountainous Areas in South China. The Qinghai–Tibet Plateau Ecological Defense shall help conserve the water sources of a few big rivers and stabilize local climate. The Loess Plateau–Sichuan–Yunnan Ecological Defense shall contribute to the ecological security in the middle and lower reaches of the Yangtze River and the Yellow River. The Northeastern Forest Belt shall protect the ecological security of the Northeast Plain. The Northern Desertification Control Belt shall form an ecological security shield for North China, Northeast China, and Northwest China. The Belt of Hills and Mountainous Areas in South China shall support the ecological security of South China and Southwest China. Except for the last belt, the Belt of Hills and Mountainous Areas in Southern, all the two defenses and the other two belts are located near or in the west of the Hu Line. The Western Development Program launched by China in 1999[8] covers one municipality (Chongqing), six provinces (Sichuan, Guizhou, Yunnan, Shaanxi, Gansu, Qinghai), and five autonomous regions (Tibet, Ningxia, Xinjiang, Inner Mongolia, and Guangxi), with a combined land area of 6.85 million square kilometers and accounting for 71.4 % of China's land area. These regions are mainly near the Hu Line or in the west of it, their ecosystems are fragile, and they lag behind other regions in economic development.

1.2 Climate Capacity

Environmental capacity is subject to influences from multiple natural factors, among which the most important one is climate conditions, especially precipitation and temperature. Even though environmental capacity, to some extent, is relative and can be changed, it is difficult to alter climate conditions with technology and economic interventions. Climate change, mainly global warming due to greenhouse gas emissions, is also a long-term and slow process involving high uncertainties. Therefore, climate capacity is a determinant for environmental capacity. Understanding the characteristics of climate capacity can help people understand, respect, and accommodate nature which can facilitate the smooth and successful transition toward an ecological civilization.

The key indicators of climate capacity are the natural levels of various climate factors, whose combinations and variations forming a region's typical climate conditions. Often, one or a few of the climate factors such as sunlight, temperature,

[7] State Council, Notice on the National Plan for Main Functions of Different Regions (No. SC [2010] 46), 21 December 2010.

[8] http://zhidao.baidu.com/link?url=KqpR2doYvnzpmetvBzW-JWJZaHWx_hTasDP3YMHR_RI6E1GQJxwHN0Iaf_eEAvQwFgiHdNVI06DA7aCKIJoAqa

and precipitation dominate or determine a specific region's climate conditions. The changes in climate factors are mainly seasonal and interannual changes. In addition, climate factors are also influenced by local topography, soil, and vegetation and can be transferred or reallocated spatially or temporally through river flow. The natural security threshold of climate capacity is subject to the constraints of extreme values of climate factors such as the precipitation level in the year of worst drought. If the interannual fluctuations or variations become bigger, it will be risky to make decisions based on multiyear average climate conditions. For instance, severe drought can lead to total crop failure, drinking water shortage, or even death among human beings and livestock.

Extreme weather events, especially extreme precipitation events, can generate serious constraints and impacts on or even lead to the collapse of socioeconomic systems and natural ecological systems. Climate factors under extreme weather events are not necessarily the thresholds of climate capacity. This is because although a region's water resource is determined by its precipitation level, but at a specific time, factors influencing the local availability of water resources include precipitation in the current year, water stored in reservoirs and ponds that is accumulated from precipitation during the previous years, as well as ground- and underground water transfer from external sources. Therefore, when assessing the threshold of a region's water capacity, the region's total precipitation and water reserve in a year or over multiple years and the supplementation of underground water need to be taken into account. Although irrigation can support agriculture production in times of drought, excessive extraction of underground water can lead to underground water depletion and reduce long-term water supply for agriculture production.

Due to topography, soil, and vegetation conditions, different regions' climate capacity can change in different directions: some regions' natural climate capacity may decline due to net capacity outflow, while other regions' may grow because of net inflow. Soil and water erosion causes reductions in local ecological capacity, while the river deltas, lakes, and other wetlands, which receive the soil and water, often witness increases in their ecological capacity. The Aihui–Tengchong Line on population density is in fact also a climate capacity dividing line based on precipitation level. Due to climate and topographic conditions, China is humid in the east and arid in the west; the altitude is high in the north and low in the south. Such climate and topographic conditions, as well as the monsoon due to air circulation, are main factors behind the existence of such a climate condition dividing line and shape the fundamental distribution of climate capacity conditions in China. The population density dividing line also reflects the fact that population distribution and socioeconomic development depend on the basic conditions of climate capacity.

Technical interventions and capital input can change the climate capacity conditions of a small area, lead to short-term changes in local climate capacity, and create the so-called derived climate capacity. Natural climate capacity is the carrying capacity of a region's natural climate conditions, while derived climate capacity is the artificial climate carrying capacity due to human social and economic activities, and it can be ecological capacity, productivity of organisms, livestock supporting capacity, or environmental capacity. They are not independent from human activities. Instead, they are derivations from the influences of socioeconomic activities on natural climate capacity. It is called derived climate capacity because despite the existence of derived climate capacity, it is still impossible for socioeconomic activities to surpass the threshold of natural climate capacity, and derived climate capacity is essentially determined by natural climate factors. Through technical interventions and scientific management, human beings can increase the ecological capacity and population supporting capacity of a region with certain climate capacity. For instance, developing crop species with high resistance to droughts and pest control can enlarge derived climate capacity under given climate conditions. Derived climate capacity mainly includes:

1. The capacity of man-made ecological systems created by such human activities as afforestation, grass planting, wetland establishment, and drinking water projects.
2. All organic materials produced and accumulated by green plants on a given piece of land during their life cycle through photosynthesis and absorption, in other words, through the transformation of matter and energy, are natural output. However, through such human interventions as plant species selection, pest control, irrigation and draining of flooded fields, fertilizer application, and hoeing as well as technology and capital investment, the land's socioeconomic output can be much higher than its natural productivity. For example, the output of agricultural products on each hectare of land, such as rice and cotton, can far exceed the productivity of natural systems under similar climate conditions.
3. Environmental capacity or carrying capacity of a given area is the pollutant and dirt absorption, damage restoration, and self-purification capabilities of water bodies, air, and other environmental media in the specific area.

Such natural capabilities can be strengthened through physical interventions, for example, spraying water and speeding up water flow to increase the total water or chemical interventions, for instance, adding detergent, in order to artificially increase the environmental capacity. For example, to a certain extent, water bodies can automatically purify themselves of such pollutants as chemical oxygen demand (COD) and ammonium; the atmosphere can also purify itself of such pollutants as sulfur dioxide and dust, provided that the pollutant concentrations are below the upper limit of allowed emissions.

It needs to be pointed out that derived capacity is a kind of artificial capacity and has three main features. First, it is subject to climate capacity and independent. Second, after the termination of human interventions, derived capacity will decline, and the relevant capacity under the given climate conditions will come back to the natural level. For example, the output and quality of farm product of a piece of abandoned cropland will fall and go back to the output level of similar local natural ecological systems. Third, the socioeconomic interventions can vary in terms of quantity, quality, approaches, timing, and duration. These are the main factors affecting artificial or derived climate capacity. If the invention is minor, for example, afforestation by aerial seeding, then the derived climate capacity will be equal to or slightly bigger than the natural climate capacity. This means the natural capacity won't decline, but its increase will be small. If the intervention is intensive, then the derived capacity can be less than the natural capacity under the given climate conditions or even negative; it can also be higher than the natural capacity or even far exceed the natural capacity. For instance, farming activities on slope fields can lead to soil and water erosion and desertification; in the end, they can change the local microclimate, make the derived capacity much smaller than the output level in the absence of human intervention, and cause negative changes in local land productivity.

The Properties of Climate Capacity Climate capacity is the natural foundation for the carrying capacity of ecosystems and the environment which provides material basis for all socioeconomic development. Climate capacity typically includes complicated ecosystems and has the following properties:

1. Rigid constraint: climate capacity within a given temporal and spatial scope is a stable natural phenomenon; it is difficult to change the rigid constraints of climate and geographic conditions with human interventions in a short period of time.
2. Variations: due to the functions and changes of climate systems, climate capacity has seasonal and interannual variations.
3. Region specific: different regions often have different climate capacity, for example, the water resource distribution of different watersheds tends to be different.
4. Conductibility/transferability: due to the influences of landform, topography, and gravity, an area's climate capacity is often closely related to the climate capacity of neighboring areas, for example, different areas in a watershed often have either water resource inflow or outflow. Moreover, long-distance water diversion projects can bring about the temporal or spatial transfer of climate capacity.
5. Interactivity/responsiveness: at the global level, the artificial socioeconomic systems and the natural climate capacity can influence each other. Human activities can change climate capacity, for instance, greenhouse gas emissions can lead to global warming and cause changes to the climate capacities of some regions. Changes in climate capacity can also affect human activities, for example, such extreme weather events as long-term drought, serious flood, typhoons, or sea level rise due to climate change can force people to evacuate and move to other regions and lead to migration.

Climate capacity is the natural capacity shaped by climate conditions. The natural output and socioeconomic development that can be supported by natural productivity are limited. If the influences of technology progress and human capital are excluded, there exists an upper limit to the socioeconomic and natural life-supporting capabilities of a given climate capacity. With population growth and the improvement of living standard, a society or an economy's demand for natural service and products will continually grow. The material output under natural climate capacity will increasingly be unable to satisfy socioeconomic demand, and the gap between them will widen (see Fig. 1.2).

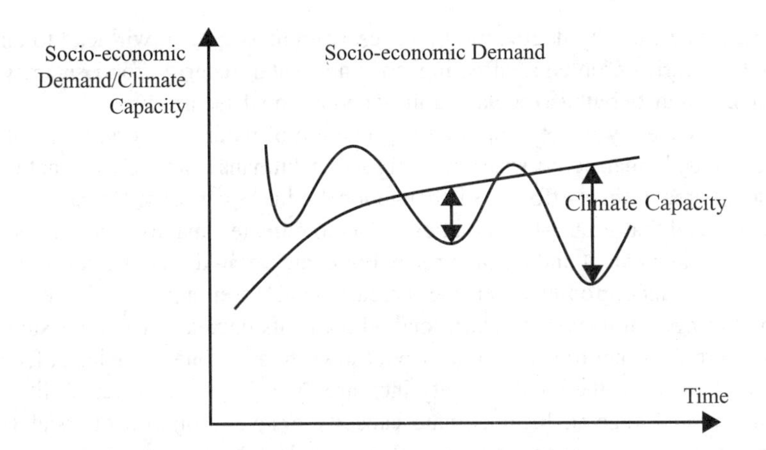

Fig. 1.2 Climate capacity and socioeconomic development

The temporal and spatial transferability of climate capacity makes it possible for societies and economic groups to use engineering interventions to improve the carrying capacity of ecosystems and the environment. Different technology choices, for example, crop selection and high-efficiency technologies, provide tools for improving ecosystem and environmental capacity from the production side. Various legal systems and policies can further strengthen the effects of engineering and technology adaptation, so as to increase the socioeconomic carrying capacity of a region under the given climate capacity. For example, greenhouse farming can increase heat supply and as a result, improve the overall production capacity inside the greenhouse. Irrigation can change water supply temporally and spatially, so as to reduce the climate capacity constraints of precipitation availability on agriculture. Reservoir construction and afforestation can conserve water resources and optimize the usage of local climate capacity.

Climate capacity is not unchangeable. Instead, it can be artificially changed through socioeconomic interventions. Generally, the following circumstances need to be noted:

1. Natural capacity: the natural security threshold of climate capacity is the natural lower limit.
2. Adapted capacity: it refers to the climate capacity increase due to climate change adaptation measures.
3. Illusionary climate capacity: it is the unsustainable climate capacity increase due to external transfer of climate capacity factors.
4. Degraded capacity: it refers to the future climate capacity degradation due to overuse and shrinkage of climate capacity factors.

When illusionary capacity occurs, climate capacity difficulties can lead to climate security problems. Despite adaptation measures, the derived climate capacity could not satisfy the socioeconomic demand; climate capacity degradation is inevitable.

When climate capacity degradation reaches a certain degree, it will lead to climate capacity security problems, affecting not only water security, food security, and economic security but also social stability and national security.

Climate capacity is not simply an aggregation of natural factors; it can also be influenced by human activities. From a big scope, humans cannot change nature, but in a small area, human efforts can change local climate capacity. However, it must be recognized that such artificial adjustments to climate capacity can only be local and to a limited extent and involve risks. For example, in dry areas, building dams or extracting underground water can increase local water supply and alleviate the water shortage. However, such artificial adjustments decrease the water supply of areas where the water transfer comes from (downstream). Due to the lack of support of natural capacity, the local capacity increase faces higher unsustainability risks, leading to shrinkage in local climate capacity or even collapse of local fragile natural systems and socioeconomic systems. Table 1.1 lists some parameters that can be used as climate capacity thresholds.

Derived climate capacity, such as environmental capacity, ecological capacity, domestic animal carrying capacity, and population supporting capacity, is essentially determined by climate factors, but technology progress and scientific management can increase them under given climate capacity. For example, in agricultural production, creating drought-resistant crop species and pest control can increase derived climate capacity, while the local natural climate capacity remains the same.

Moreover, through the interactions between natural systems and environmental systems, climate capacity can be created, transferred, and changed. Natural forces mainly include water systems in a watershed which gathers precipitation of the upstream areas and through water flow increases the water capacity of the middle stream and downstream areas; they can also be snowfalls on high mountains: the precipitation is stored in the form of snow and ice in cold winter and melts in summer and autumn and increases water supply. In this way, it realizes the temporal and spatial transfer of water supply. In human history, there have been examples of artificial adjustments to climate capacity, for instance, changing natural ecological environments into areas suitable for human settlement or the transfer of ecological footprint through production and trade. The first category mainly consists of engineering measures, including artificial influences on weather, water diversion projects, irrigation facilities, ecological protection, and so on. The second category includes the import and export of grains, woods, and energy-intensive products—such trade essentially realized the temporal and spatial transfer of embedded energy and embedded water resources. Fossil fuel, the energy foundation for industrial era, in fact comes from solar energy stored in dead organisms millions of years ago. It is another example of the transtemporal and spatial utilization of climate capacity.

When performing temporal and spatial adjustments to the climate capacity of a certain region to meet the demand of human societies, economies, and natural systems, various social, economic, and natural system factors need to be taken into account. The principles to be followed mainly include:

Table 1.1 Thresholds and parameters for climate capacity

Climate and its derived capacity thresholds	Threshold factor	Parameter (threshold)	Policy approaches for capacity increase or stabilization
Climate capacity threshold	Temperature	Accumulated temperature[a]	Greenhouse
	Precipitation	Average annual precipitation	Increasing the coverage rate of green plants and grass, protecting and expanding wetland area
Derived capacity: ecological threshold	Carrying capacity of water resources	Per capita useable water resources (≤500 m^3)[b]	Changing the temporal and spatial availability of water resources through engineering measures
	Carrying capacity of ecological systems	Quantity of animal (e.g., livestock) and plant biodiversity (high, medium, low)	Reducing human interventions
	Carrying capacity of land	Per capita useable land area; productivity of per hectare of land	Using irrigation facilities to improve water supply, applying fertilizer, using improved species, and pest control
Derived capacity: environmental capacity	Chemical oxygen demand (OCD)	Quality standards for drinking water	Engineering, technical, and institutional measures
	PM2.5 concentration in the air	50 ug/m^3	Reducing the combustion of fossil fuel, tightening emission standards, increasing forest coverage, and spraying water to reduce dust in the air

[a]Accumulated temperature is a concept often used in agriculture. It refers to minimum temperature (heat) conditions during a crop's entire life cycle and is the sum of daily average temperatures. Agriculture production is restricted by a specific area's average temperature and the accumulated temperature a crop requires

[b]According to the World Bank standards, the definition of water capacity is based on per capita water resource possession: slight scarcity is ≤3000 m^3/person; medium scarcity is ≤2000 m^3/person; severe scarcity ≤1000 m^3/person; and extreme scarcity ≤500 m^3/person. China's per capita water possession is 2240 m^3, ranking the 88th in the world. The geographic distribution of China's water resource is highly uneven. The Yangtze River shed and south of it covers 36.5 % of China's land area but has 81 % of the water resources nationwide. In the north of the Yangtze River, there is 63.5 % of China's land, but the water resources are only 19 % of the national total. Currently, 16 of China's 32 provinces, autonomous regions, and municipalities are below the threshold of severe water scarcity. Among them, five provinces (Hebei, Shandong, Henan, Shanxi, and Jiangsu) and one autonomous region (Ningxia) find their per capita water resource possessions are less than 500 cubic meters (Source: World Bank 2011, *Addressing China's Water Scarcity—Recommendations for Selected Water Resource Management Issues*)

1. The principle of economic rationality. In the course of designing measures to change the climate capacity of a region, economic cost and benefit needs to be considered. For example, people can imagine such dramatic measures as diverting water from the Bohai Sea to Inner Mongolia and cutting through the Himalaya Mountains. Such nature-transforming measures are in line with the values of industrial civilization and are technically feasible, but they are not economically feasible.
2. The principle of ecological environmental integrity. The impacts on the regional and wider ecological environments need to be taken into account when deciding climate capacity-changing measures. Due to its rigidness, climate capacity could not be changed at the global level. Temporal and special adjustments to local climate capacity involve the transfer of climate factors and natural systems.
3. The principle of protection. From the perspective of climate security and protection needs, adaptation measures should give priority to the regions and social groups with poor basic living conditions and vulnerable to property losses and casualties in case of climate disasters.
4. The principle of equitable distribution. The change and transfer of climate capacity are in fact a redistribution of climate resources, and priority should be given to the most vulnerable social groups and the most pressing needs for fairness in resource distribution and benefit sharing.

The above principles can be of different importance under different circumstances. For example, if migration and disaster relief is needed for property losses and human casualties due to sea level rise and climate disasters, the climate protection principle should be of the highest priority, and the economic costs should be a less important consideration.

1.3 Climate Migration

Large-scale population migration in human history has been in most cases due to the fact that the social and economic needs of the population in a certain region exceeded the local climate carrying capacity. In Western China, climate carrying capacity, especially water resource carrying capacity, is low; water resources, which are indispensable for natural ecological system and human social and economic activities, are extremely scarce. In industrial civilization, technology and capital inputs can increase the derived climate capacity to some extent, such as ecological carrying capacity. However, the rigidness of climate capacity sometimes leads to spontaneous or organized population migration. Such migration due to climate carrying capacity overpassing is different from the migration of agricultural population into cities. The latter is not because human activities exceed local ecological carrying capacity. Instead, it is a voluntary migration because of economic purposes.

In China, poor regions are mainly in the western plateau and the mountainous and hilly regions in Central and Western China, including most parts of Tibet,

Xinjiang, Qinghai, Gansu, Ningxia, Shaanxi, Sichuan, Guizhou, Yunnan, and Guangxi as well as Inner Mongolia and the Yimeng Mountain area, Taihang Mountain area, Lvliang Mountain area, Qinling–Daba Mountain area, Wuling Mountain area, Dabie Mountain area, Jinggang Mountain area, as well as the southern part of Jiangxi Province. These regions are of fragile ecological systems and low economic development level. Many of these regions use migration as a way of poverty alleviation and call it ecological migration. Migrating people from the mountainous areas in South China can in fact benefit the protection of local ecosystems. However, for the arid and semiarid regions in Northwest China, where local climate capacity is unable to support local population, the causes and nature of migration are different. For example, since the 1980s, Ningxia has migrated over 600,000 people in its central and southern arid parts; the autonomous region plans to resettle another 350,000 people during the 2010–2015 period. The organized migration in Ningxia is aimed at poverty alleviation, economic development, and ecosystem protection. On the face of it, the migration is due to ecological environmental deterioration and poverty, but the driving force behind the vicious circle of population stress–ecosystem deterioration–poverty is climate change. Therefore, the migration in Ningxia is essentially "climate migration"—it is an organized resettlement of a large number of people because the arid environment and lack of precipitation make some areas unsuitable for human settlement, and migration is an adaptation measure in response to current or long-term unfavorable climate change. Fundamentally, the development of these regions is restricted by local climate capacity; local natural ecosystems cannot provide sufficient produce and have low population carrying capacity. Therefore, the part of population which exceeds local climate carrying capacity has to move to other regions. Therefore, climate migration pays more attention to the purposes of climate protection and security, and it is essentially different from migration aimed at ecosystem protection and economic development.

The migration in Xihaigu in Ningxia provides a typical case for analyzing the relationships between ecological capacity and development from the perspective of climate capacity.

"A region's natural environment is unable to support the local people." Climate capacity restricts the population carrying capacity of Xihaigu Region in Ningxia. Xihaigu, which is known as "with the worst natural environment in China," is one of the regions subject to key national support for poverty alleviation. It includes two districts (the Yuanzhou District and the Hongbaoshi District) and seven counties (Xiji County, Longde County, Jingyuan County, Pengyang County, Haiyuan County, Tongxin County, and Yanchi County). Together, they account for about 60 % of the land area in Ningxia, and their combined population is approximately two million, accounting for one third of the population in Ningxia. They are also the biggest concentrated area of the Hui ethnic group. They are located at the transition area from semiarid Loess Plateau toward the arid and windy desert, and the local economy is a mixture of farming and animal herding. Local ecosystems are fragile, are dry, and lack of precipitation, and the soil is futile. The local resource endowment is poor, and natural disasters are frequent. The area is plagued by soil and water erosion. The annual average precipitation is 200–650 mm, and the per capita water resource possession is only 136.5 m^3. It is one of the regions with extreme water scarcity in China. In this region, almost one million people are still living below the Chinese national poverty line of per capita income 1350 RMB in 2010. Among them, 350,000 people are living in arid mountain areas far away from main roads, with little access to information, few nonagricultural job opportunities, poor ecological systems, serious water scarcity, and very poor natural conditions.[9]

Most of the farmers included in the migration plan of Guyuan City live in areas with high mountains and deep valleys; their livelihood is badly restricted by local climate capacity, and the farmers struggle at a substance level. Guyuan City's experiences with agricultural and industrial development and city construction indicate that for this region, economic development cannot lift local poor people out of poverty; on the contrary, economic development deteriorates poverty because local ecosystems are too fragile. In fact, fragile ecosystems reflect the constraints of climate capacity. The region's problem cannot be solved through regular economic development or poverty alleviation-oriented development. Economic development (more investments in infrastructure, water resource exploration, developing modern industry, and urbanization) can only further deteriorate local ecological environment and cannot increase local climate capacity and climate adaptation capacity.

Climate capacity is not only an issue faced by regions with people moving out; it is also a problem for regions which accept migrants. Due to the restrictions of water and land resources, among the migration organized by the Ningxia Hui Autonomous Region government, only a small share of migrants were moved to the areas with irrigation networks from the Yellow River, where the climate capacity has been increased, and they are able to support higher population density. Many migrant-receiving areas are of low climate capacity; after accepting immigrants, the larger population's demand for water resource further increases, and the extraction of underground water grows. As a result, their fragile climate capacity deteriorates due to higher population pressure. For instance, Hongsibao District, which is located in the irrigated area in central Ningxia and the biggest migrant-receiving area, was originally desolate sands. With the migration policy support, irrigation networks have been constructed to divert water from the Yellow River for irrigation, and the desolate sands have changed into an agricultural oasis. It received 190,000 migrants and became a good migrant resettlement example in Ningxia, even the entire country. Because of the successful example, the area has also attracted an endless flow of spontaneous migrants from such poor areas as south of Ningxia as well as the neighboring Shaanxi and Gansu. Hongsibao is a typical example of climate capacity transfer through human intervention; in other words, it utilizes the water transfer from the upper stream of the Yellow River and reduces the water available to the lower reaches of the Yellow River. However, in the long term, the area is still of fragile climate capacity and unstable. In fact, the local government has realized that the successful migration and population concentration effect provides development opportunities for the area but also highlights the water resource and environmental

[9] See the *Ecological Mitigation Plan for the 2011–2015 Period in Central and Southern Parts of Ningxia* by the Ningxia Development and Reform Commission.

constraints to the development of the newly urbanized Hongsibao District. Due to water shortage in the middle and lower reaches of the Yellow River, the Yellow River Conservancy Commission has allocated water utilization quota for the provinces and autonomous regions along the river. A region's quota-exceeding use of water from the Yellow River will inevitably lead to declines in water environmental capacity of the lower reaches. In fact, the drying up of the Yellow River in the lower reaches, to some extent, is due to the water diversion for irrigation purpose in the upper stream. Moreover, in case of major fluctuations in the water flow of the Yellow River because of climate change in the future, the population and development pressure faced by the "desert oasis" Hongsibao will intensify and even threaten its existence.

Climate migration can happen in three different circumstances. The first circumstance is that the climate capacity declines because of natural or human factors, but the human social and economic activities do not change much. The second circumstance is that the climate capacity remains the same, but the human social and economic activities continually increase and exceed the climate capacity. The third circumstance is the combination of climate capacity shrinkage and human activity expansion. Under the background of climate change, the third circumstance will become more common. Due to irreversible or abrupt extraordinary change of one or more climate and ecological factors (especially temperature and precipitation), the climate capacity declines and can no longer support the existing population size, economic activity type, and intensity. Under such situations, further population increase and higher social and economic activity intensity lead to the occurrence of the vicious circle of environmental degradation–poverty or loss of livelihood for a short period. To adapt to the impacts of climate change, people spontaneously or systematically migrate to other places temporally or permanently. Since abrupt climate changes are often temporal and reversible, such climate migrants are mainly emergent climate refugees, while climate migration due to irreversible and sustained changes is foreseeable and permanent. Although climate refugees can also become permanent migrants, climate migration is often permanent and irreversible.

The international community divides migration due to climate change into several major types: migration caused by emergent climate disasters (e.g., typhoons and floods), migration due to gradual climate disasters (for instance, sea level rise and salinization), migration from small island countries, migration from high-risk areas, as well as migration due to resource and political conflicts. Climate migrants usually have to leave their original living places due to threats to their livelihood, life, and properties as well as living environment due to emergent climate disasters (such as typhoon and flood), long-term climate risks (for instance, sea level rise), or gradual ecological environmental changes (e.g., droughts). Although it is difficult to forecast the occurring places and moving directions of climate migration, it is certain that regions with high climate disaster risks and ecological environment-sensitive areas are most vulnerable to climate change and are often areas with high climate migration occurrence. They include cities on river delta, small island countries, low-lying areas, arid areas, polar regions, as well as regions prone to extreme climate events.

The implications and characteristics of climate migration can be analyzed in depth from the causes and purposes of climate migration, the principles or considerations of policy intervention, and the governance bodies as well as approaches. For example, the following analysis is made on ecological migration:

The Causes of Climate Migration Climate migration occurs because climate capacity constraints are rigid, and the population size and local social and economic activity intensity exceed the local climate carrying capacity. Due to long-term climate change, ecosystems and human settlement environment change over time and are unfavorable to human settlement and life. On the surface, the ecological migration policy implemented in Western China is triggered by ecological environmental deterioration and poverty, but the driving factor behind the vicious circle of population pressure–ecological degradation–poverty is the rigid constraints of climate capacity. Human social and economic activities cannot expand or increase climate capacity; therefore, they have to be reduced to decrease the pressure on natural environment through migration. Unlike such involuntary climate migration, "ecological migration" emphasizes the purposes of protecting ecosystem service and biodiversity through such ecological protection measures as establishing natural protection zones, converting farmland into lakes, converting grassland into forest, and converting degraded farmland into grassland. The migration under such projects is often human resettlement organized by local governments, involuntarily supported by relevant economic compensation. The cause of migration can be that the population of a specific region exceeds the carrying capacity of local ecosystems, and it is necessary to restore the health of local ecosystems as soon as possible, for example, converting farmland to forest, to lake, or to grassland. They can happen not because the emigration area is unable to support human settlement but in order to protect certain species and ecological value. Examples are waterhead area protection and panda habitat protection. For instance, the ecological migration in the mountain areas in Zhejiang Province is not because the mountain areas are unable to support local people's livelihood; it is more for economic interests. In order to protect the ecosystems on the mountains, the government helps the people move to plain areas, provides better education for their children, and helps them find jobs in the plain areas so that they do not return to the mountains. The ecological migration from the mountain areas of Zhejiang is not forced migration due to environmental pressure. Instead, it is voluntary and compensated migration. It is an active retreat, not a forced migration, because fragile ecosystems are unable to support human livelihood.

The Purposes of Climate Migration The migration decision is often for multiple purposes, for example, seeking safety, higher income, and better living conditions. To distinguish climate migration and ecological migration, it is necessary to analyze the main purposes and secondary objectives of a specific migration. The main purposes of climate migration are for the migrants' substance and to make the population size and the intensity of social and economic activities in line with local climate capacity, so as to achieve population security and environmental sustainable development of the specific region. Although climate migration can generate some side effects of ecosystem restoration, the main purpose is not the reestablishment of ecological functions. In contrast, the primary objectives of ecological migration are ecological environmental protection and ecological service restoration, but in practice, ecological migration can also combine such secondary objectives as to lift the migrants out of poverty, change their lifestyle and production approaches, and realize coordinated development of regional economy or even the future survival and development space of local societies. This leads to vagueness in the policy objectives of ecological migration and many problems in practice and policy implementation effects. Therefore, it is necessary to differentiate different types of migration and tailor the policy design accordingly, so as to make migration decisions based on specific situations.

The Policy Foundation of Climate Migration Climate change is essentially an externality problem associated with global environmental public good. Compensations to climate migrations should follow the principle of climate security and focus on the primary task of securing basic development needs (poverty alleviation) and climate protection. At the same time, they should also take into account the climate justice principle and the principle of giving priority to regions highly vulnerable to climate change and weak social groups. Ecological migration should follow the principle of "protectors enjoy the benefits." Therefore, there are differences between climate migration and ecological migration in migrant compensation, funding sources, and policy implementation authorities. Unlike climate migration, ecological migration is primarily aimed at restoring the ecological environment of ecological fragile areas. Therefore, ecological migration should follow the principle of ecological compensation, and the compensation should be based on "payment for ecological service" or PES. It is necessary to evaluate the ecological service benefits provided by the protection zones due to ecological migration such as prevention of soil and water erosion and biodiversity protection. Such benefits can be evaluated and quantified using relevant methodologies in environmental economics and based on market values. Currently, some regions in China estimate the compensation for ecological migration based on the payments for ecological services. For example, Zhejiang's compensation to the ecological migration for the conservation of drinking water sources is based on the payment from Shanghai for protecting the quality of Shanghai's drinking water sources. The amount is estimated based on drinking water quality levels and public benefits of improved ecosystem, consumers' willingness to pay, and payment capability, and the principle is that the beneficiaries should pay for the ecological services and compensate the ecological migrants.

Which one to choose, ecological migration or climate migration? In face of environmental and climate change, people may take three different attitudes or reactions. The first is passively accepting unfavorable impacts; the second is actively seeking to reduce the impacts; the third one is retreating from the impacted areas. As the last option, migration has the negative side effects of possibly increasing the resource (such as food supply) and environmental pressures and conflicts in the migrant-receiving regions. There are two different opinions on migration due to environmental and climate change. Traditionally, people think migration is a kind of failure, and it indicates that the local residents are unable to adapt to environmental deterioration and climate change. This is reflected in the concept of climate refugees. A new opinion that is gradually accepted in recent years regards population resettlement as a response to environmental and climate change. Since the 1980s and the 1990s, China has implemented many ecological migration projects in the ecological fragile and poor regions in Western China and the regions prone to natural disasters in the Yangtze River basin. These ecological migration projects are closely related to environmental and climate change. In fact, the ecological migration organized by local governments in many parts of China is active and planned adaptation to environmental and climate change.

It is apparent that ecological migration shows remarkable regional differences. Ecological is more typical and common in regions with poor geographic and climate conditions, fragile ecological environment, and low population carrying capacity. According to studies on poverty in China, resource scarcity, bad ecological environment, and natural disasters are main causes of poverty. The majority of poverty regions in China are in areas subject to major global climate change impacts. The distribution of poor population is highly consistent with the distribution of fragile ecological environmental areas. Among the population living in ecological sensitive areas, 74 % are in poverty counties, accounting for 81 % of the total population living under the national poverty line. Ecological poverty and climate poverty caused by ecological environmental deterioration have become a regional feature of poverty in Western China. Therefore, the migration conducted in Western China is close to the definition of climate migration, and the factors behind it are "climate capacity constraints" and "poverty trap." The rigidness of climate capacity restricts the development space and the choice of development approaches; even the social and economic demand of the small population cannot be satisfied, and the local people struggle in poverty and find it difficult to fundamentally get rid of poverty. These problems are closely related to the vulnerability due to climate change.

In Western China, the annual precipitation is low, and the effects of ecological migration are not as good as expected, partially because of the vagueness of ecological migration purposes. In practice, ecological migration is often considered a low-cost and high-benefit way of alleviating population pressure on land carrying capacity and solving the conflicts between ecological environmental protection and lifting local farmers and herd population from poverty. However, in some Chinese regions, ecological migration is often expected to realize multiple objectives such as ecosystem restoration, poverty alleviation, and economic development which complicate the concept of ecological migration. Due to over aggregation and generalization of the ecological migration concept, the different environmental driving forces and objectives are often blurred, making the concept lack of solid theoretic basis and clear objective orientation in practice. As a result, policy designs, compensation standards, and migration approaches vary in different regions, making it difficult for policy effect evaluation and practice improvement.

The facts of forced population migration due to climate capacity constraints indicate that the technological and capital interventions in industrial civilization cannot improve or increase the natural climate capacity. Instead, human beings have to respect the capacity constraints and through population resettlement, keep human activities within the threshold of climate capacity.

1.4 Allowing Nature to Take Its Course

The shared values of industrial civilization are about transforming nature through human activities. However, due to such natural factors as natural climate conditions and topography, human beings have not changed nature; moreover, the regional differences in economic development further highlight the importance of natural conditions. It is necessary for mankind to review the theories and practice of industrial civilization and change their shared values from transforming nature to respecting nature.

A precondition of realizing the Beautiful Chinese Dream is that China's social and economic activities must not exceed the carrying capacity of ecosystems. The resource and environmental capacity of the planet Earth is limited; in other words, the ecological supply is limited. If the demand for ecological service exceeds the supply capacity of ecosystems on the Earth, ecological degradation is inevitable, and the beauty of nature will be damaged. To build a Beautiful China, it is necessary to balance population, resource, and environment, harmonize economic benefits and social and ecological benefits, and respect, accommodate, and protect nature according to the shared values and principles of ecological civilization. To realize the magnificent blueprint of Beautiful China, it is not only necessary to keep natural supply of ecosystems at optimal level, but more importantly, people have to control the demand on ecosystems, so as to keep human's ecological footprint below ecological carrying capacity and maintain ecological security.

To Change Consumption Patterns and to Reduce Ecological Footprint Ecological footprint refers to the land (or waterbody) area with ecosystem productivity that is needed for providing the various commodities and services directly or indirectly supplied by natural ecological systems for human consumption and absorbing the wastes generated during the commodity and service production and consumption.[10] Therefore, ecological footprint is the supply and demand relations between the ecological demand (or ecological consumption) and the productivity (or carrying capacity) of ecosystems. Ecological footprint and carrying capacity are measured in "global hectare." One global hectare is one hectare of land use based on global average ecosystem productivity. The ecological footprints of some key ecological factors, such as greenhouse gas emissions and water resource consumption, are expressed in carbon footprint and water footprint, and their units

[10] The concept of ecological footprint was first formulated by William Rees in 1992. See Rees, W. E. (1992, October). Ecological footprints and appropriated carrying capacity: What urban economics leaves out. *Environment and Urbanisation*, *4*(2), 121–130.

are also land area based on global average ecosystem productivity. Ecological footprint is an effective tool for measuring human beings' demand for and consumption of natural resources. It quantifies the supply of and the demand for renewable resources from ecosystems and provides a rational basis for the environmental and economic policy making and the selection of production and consumption patterns and offers an objective criterion for ecological civilization building.

The supply of ecosystems is subject to natural capacity constraint despite the fact that ecosystems' carrying capacity can, to some extent, be increased through investments and technical interventions. However, the natural capacity itself cannot be fundamentally changed. Ecological footprint is different, and it examines the demand of human consumption for ecosystems. For example, the ecological footprint of human food consumption depends on the contents of consumption. Grain production needs certain size of arable land; if the consumption is animal-based food such as beef and lamb, it needs more land and has bigger ecological footprint than grain production. Industrial products and infrastructure investments are also directly or indirectly for human consumption. The ecological footprint of different consumption choices, such as taking public transport or driving a car, can also be many times different. The ecological footprint of different consumptions can be calculated. For example, steel production uses energy, discharges pollutants, occupies land, and consumes water. Based on the amount of energy each hectare of land can fix and convert and the amount of land use, the global hectares of land for steel production and consumption can be calculated. The ecological footprint of the emissions of carbon dioxide, a kind of greenhouse, can also be quantified based on the rate of green plants that fix and absorb carbon through photosynthesis.

No matter how advanced science and technology is, human beings always need to rely on the nature for water, food, and energy supply. Since the 1970s, human beings' annual demand for the ecosystems on Earth has exceeded their regeneration capability. According to the estimate data in 2012,[11] the world ecological footprint in 2008 was 18.2 billion global hectares and 2.7 global hectares per person. In the same year, the carrying capacity of ecosystems worldwide was 12.0 billion global hectares or 1.8 global hectares per person. This means that in 2008, the global ecological deficit was as high as 50 % and mankind needs one and a half Earth to produce the renewable resources for human consumption and to absorb the carbon dioxide emissions from human activities. If such a trend continues, by 2030 even two Earths won't be enough to support mankind's consumption demand. China's fragile ecosystems are under huge and constantly increasing population and economic growth pressure. Estimates by the World Wildlife Foundation (WWF)[12] indicated that in 2008, China's per capita ecological footprint was 2.1 global hectares per person which was around 80 % of the global average. However, China's ecosystems are fragile, and over half of the country's territory is plateaus, mountainous areas,

---

[11] The WWF organized Chinese and international experts to estimate the ecological footprint of China and released the results in 2002. See WWF *China Ecological Footprint Report 2012*. Beijing. 64 pages (www.wwfchina.org).

[12] WWF. *China Ecological Footprint Report* 2012. Beijing. www.wwfchina.org

deserts, and Gobi deserts, and the productivity of Chinese ecosystems is far below the global average. China's ecological footprint has exceeded twice the productivity of the country's ecosystems. Since the turn of the century, China's annual imports of such natural resources as oil and iron ore have been growing at high speed which is another indicator of the country's ecological footprint deficit. Moreover, China's target of creating a well-off society by 2020 means higher urbanization rate and higher-quality consumption in the years to come. The existing lifestyle and production patterns, the ecosystems which have been used beyond its carrying capacity, and the development based on ecological deficit have threatened and been damaging ecosystem security which is the foundation of social and economic development.

Building ecological civilization is to consciously select a consumption pattern to reduce the human being's ecological footprint while securing the quality of human life. For example, the income and living standard of the United States and the European Union is more or less the same, while the per capita carbon dioxide emissions of the United States were 2.4 times the level in the European Union.[13] Such differences are because of the different lifestyles. To build a Beautiful China, the Chinese people must adjust and change their consumption patterns and reduce the country's ecological footprint, so that the country's demand for ecosystems can be below the ecosystems' carrying capacity and maintain the natural foundation of "Beautiful China."

Allowing Nature to Take Its Course and Maintain the Productivity of Ecosystems If the productivity of ecosystems were damaged by human activities and the ecosystems are full of scars and wounds, the objective of creating a "Beautiful China" cannot be realized. Natural productivity is the foundation of beauty; nature will show its beauty if it is not under pressure and left to take its own course. To maintain the productivity of ecosystems, human beings need to follow the shared values of ecological civilization and allow nature to take its course instead of working against nature and damaging nature under the name of transforming nature.

In the 1980s, at the beginning of large-scale industrialization and urbanization after the reform and opening up, China clearly formulated the policy of protecting "ecological balances." The ecological balance policy was mainly aimed at maintaining the stability of ecological supply through protecting nature. After the foundation of the People's Republic of China in 1949, social order was restored, the country's economy kept growing, and the population increased quickly. These factors requested more ecological supply. The nationwide campaign of "learning from Dazhai in agricultural production" launched in the 1960s included reclaiming land from lakes and slashing and burning forest for land. It was aimed at getting more output from ecosystems and expanding arable land. However, it caused soil and water erosion, blocking of river courses, and frequent droughts and floods. In many

[13] The per capita carbon emissions of the European Union and the United States have passed their peak, and the majority either decline or stabilize. Since 2000, the EU's per capita emissions have declined a lot. The data here are for 2011. See BP Statistical Review of World Energy 2012. The World Bank WDI database (http://data.worldbank.org/data-catalog).

cases, the efforts not only failed to increase the outputs of ecosystems but also led to ecosystem degradation and declines in ecosystem productivity. Therefore, the ecological balances in the 1980s were primarily aimed at restoring nature, and environmental pollution was not a main threat to ecosystems during that period.

Since the late 1970s, China's rapid industrialization led to the flow of investment and employment to the manufacturing sector, minerals and fossil fuels were explored and used in large scale, and large quantity of industrial waste entered the environment. The damages to nature include productivity degradation of ecosystems and poisonous damages to some ecosystems. The productivity of agriculture has been improved, the output of each hectare of land is higher, and the variety of commodities is much more diversified. However, China's water is polluted, and its air is full of pollutants. The quantity of food supply is high and stable. However, food safety problems are widespread. Heavy metal pollution to soil, pesticide residues in food and soil, and particulate matters (PM10 and PM2.5)[14] in the air not only affect the quantity of ecosystems' material output; they also influence and poison the quality of ecosystems and their products. The results are damages to both the health of ecosystems and human health. Compared with ecosystem degradation, ecosystem poisoning causes even longer-term damages to ecological security.

Respecting nature is a kind of ethic value; the rules for human behavior should be allowing the nature to take its course. If people do not follow natural rules, it is violating the principle of respecting nature and leads to damages to ecosystem productivity. The key of allowing nature to take its course is keeping human activities within the carrying capacity of ecosystems. Building ecological civilization is essentially building a resource-saving and environment-friendly society that is based on resource and environmental carrying capacity, follows the laws of nature, and aims at sustainable development, so as to maintain the productivity of ecosystems. The ultimate constraints of ecosystem carrying capacity are reflected in ecosystem productivity. China has a vast territory; the remarkable different climate conditions of different regions and the distinctive spatial differences in ecosystem productivity shape the country's corresponding social and economic activity distribution. The environmental carrying capacity is low, and the ecosystem productivity is weak in Western China; therefore, they cannot support large-scale urbanization and industrialization. The productivity of ecosystems on the same size of land in East China can be several times or higher than the level in Western China. Allowing nature to follow its course does not mean making all regions reach the same level of social and economic development. Instead, it is making local development in line with the productivity of local ecosystems and local resource and environmental carrying capacity.

[14] PM is particulate matter. PM10 and PM2.5 are particulate matters with diameters below 10 μm and 2.5 μm. PM10 can enter people's breathing systems and PM2.5 can enter people's lungs. As the particulate matters often contain chemicals and heavy metals, they are important disease-causing factors.

Respecting Nature and Making Human Activities in Line with the Carrying Capacity of Ecosystems Can natural beauty be produced or created through technology innovation and capital investment? Ecosystems have their inherent spatial structure and temporal changes. Can the productivity of the part of an ecosystem or an ecosystem be increased through changing the natural structure with engineering and technological intervention?

In industrial civilization, people believe that technology and investment can increase environmental capacity and improve ecosystem productivity. However, the partial improvement of a small ecosystem does not mean the enlargement of natural capacity. Moreover, the appearance of capacity expansion through technology and investment can artificially increase the risks and vulnerability of ecosystems. Bigger investments can lead to higher risks. For example, some projects of diverting water from the Yellow River for irrigation in the lower reaches of the Yellow River can artificially increase the ecosystem productivity of the irrigated areas. However, once the Yellow River dries up seasonally, the environmental capacity and carrying capacity of the irrigated area or cities will disappear. Additionally, diverting the Yellow River water for irrigation is to some extent a zero-sum game. One area's diverting water for local irrigation reduces the water available for other areas because the precipitation at the origin or the watershed of the Yellow River is a rigid climate capacity and the water resource accumulated is fixed. If it exceeds the capacity of natural ecosystems, increases in the water resource allocation to one area inevitably lead to decreases in water resource for other areas. Another example is the water management of Beijing. Due to the low precipitation of North China, Beijing faces the problem of water shortage. To guarantee water supply in Beijing, the water use in areas around Beijing is restricted. This can significantly increase the economic and social benefits of water use. However, it is a typical "zero-sum arrangement" in water environmental capacity allocation. The middle route of the "diverting water from the south to the north" project is using the water from the Hanshui River watershed to increase water supply in Beijing, so as to increase the ecosystem carrying capacity of China's capital. However, the 1200 km of long-distance water diversion depends on the water availability and quality of the Hanshui River watershed. In case the water quantity and quality of the Hanshui River changes, the imported ecosystem carrying capacity of Beijing will face high risks and vulnerabilities. Under industrial civilization, there is the practice of using technological interventions and social governance to restrict one area's beauty or use one region's resources of beauty to improve or decorate the beauty of another region. If such practices exceed certain limit, it is disrespecting nature, and the transferred beauty is at the price of another region's loss.

Obviously, the technologies of transforming and conquering nature based on the shared values of industrial civilization are not completely in line with the needs of building ecological civilization and Beautiful China. The principles of ecological civilization are respecting nature and following the laws of nature. The technologies needed in ecological civilization are based on the precondition of respecting ecosystem carrying capacity. For example, energy-efficiency technologies and renewable energy technologies do not generate "zero-sum" effects on ecosystem carrying capacity. Instead, they can achieve real enlargement to climate capacity or increases to ecosystem carrying capacity. Of course, energy-efficiency increase has its limit, and renewable energy production is also subject to constraints. For example, solar heating or solar photovoltaics (pv) utilization is subject to the constraint of total solar radiation on the surface of Earth. There is no technology to increase the surface area of Earth. However, through technology innovation, human beings can increase the utilization efficiency of the limited solar light resources. This means some technologies under the shared values of the industrial civilization are compatible with ecological civilization building and some can be upgraded and transformed by the ecological civilization beliefs. Ecological civilization building and Beautiful China do not require full load functioning of ecosystem capacity. Instead, there should be some idle space for use by other biological communities; part of the ecosystem capacity should also be reserved for future generations.

Therefore, economic development should be based on respecting nature. The increased carrying capacity obtained through conquering and transforming nature needs to be assessed based on the productivity of entire ecosystems. During the utilization of ecosystem carrying capacity, the transfer payment or price of ecosystems needs to be taken into account, so as to base human activities on the scientific knowledge on nature and respect of nature.

Protect Nature and Improve Ecological Security Building "Beautiful China" needs to develop economy and improve environment on the basis of existing technical and economic conditions and according to the spatial distribution of ecosystem capacity. For areas where the intensity of social and economic activities has already exceeded the carrying capacity of ecosystems, the problem shall be addressed through three aspects: reducing ecological footprint, gradually allowing nature to take its course, and aligning human activities with ecosystem carrying capacity.

First of all, the intensity and level of social and economic activities should be properly reduced. Converting farmland back to lakes, converting farmland back to forest, and converting farmland back to grassland are the most effective measures. However, they are also likely subject to the biggest constraints or resistance. This is because the demand of a region's social and economic activities for ecosystem output or ecological footprint is also to some extent rigid. For example, the limited climate capacity of Xihaigu in Ningxia is unable to support local population, and the resettlement of the excessive population needs to be supported by corresponding climate capacity. The water use in Beijing surpasses the local ecosystem carrying capacity; the first solution is to move some water-intensive industries out of Beijing. However, this could have negative impacts on Beijing's local Fiscal income and employment. Second, efforts should be made to upgrade the technical level and improve the policy mechanisms. Higher resource utilization efficiency means double or more output with the same resource input. Creating and spreading the use of drought-resistant crop species in arid areas or improving productivity of land can increase grain produce and satisfy social demand with the same or even less pressure on resources and environment. Third, leave space for natural ecosystems to restore themselves. The long periods of urbanization and industrialization have consumed lots of natural resources and caused many damages. It is already difficult for natural ecosystems to meet the demand of human activities, leading to serious ecosystem degradation in many places. To protect nature, "restoration" activities in violation of natural laws must be avoided. In regions short of water, obviously environmental beautification should avoid water-intensive lawns. Planting trees in arid areas which are only suitable for herb growth is in fact damaging the environment.

Building "Beautiful China" must accommodate nature, so as to make the beauty real and sustainable. Without the relevant ecosystem carrying capacity support, artificial ecological security seems to protect nature, but in fact it causes damages to nature. Many cities in arid and semiarid areas build large artificial lakes to "beautify" the local environment which are expensive, waste precious water resources, and work against nature forces. Moreover, such "beauty" cannot be sustained. Many local Chinese governments build super large squares and excessively wide roads at the price of occupying valuable ecosystem capacity space and damaging natural beauty. It should be noted that artificial buildings, to a large extent, are irreversible. Soil that can conserve water and is suitable for life survival is formed through thousands of years of natural process; reinforced concrete, once built and become a landscape, will take at least hundreds of years to decompose and recover the productivity of natural ecosystems.

Chapter 2
The Development Paradigm of Ecological Civilization

The material wealth accumulated in the last 300 years since industrial revolution has far exceeded the total quantity of material wealth human beings created before that. With industrialization—the new human social civilization form—industrial civilization emerged, developed, became mature, and gradually replaced the agricultural civilization of low productivity. People's shared values, production and life patterns, social organization forms, and institutional and legal systems also changed dramatically in the process. However, while people create and enjoy abundant material wealth, their living environments have been deteriorating: the fresh air becomes filthy, clean water source is polluted, agricultural products from soil polluted by heavy metal are poisonous and unhealthy, the global climate is becoming warmer, the ecosystems are degrading, and the income gap between the rich and the poor is widening. People have to rethink: can we continue the development paradigm of industrial civilization? If a new development paradigm is needed, then what shall it look like? In over three decades since the reform and opening up, China has experienced rapid industrialization and urbanization, and the development paradigm of industrial civilization has become dominant. With the rapid economic growth, the problems of imbalance, inharmony, and unsustainability are becoming increasingly severe. Under such a background, China's traditional, historical, and cultural concept and practice of "harmony between human and nature" is passed on, promoted, and refreshed, and the new development paradigm of ecological civilization is formed and evolved.

2.1 Criticism on Industrial Civilization

The success of industrialization made industrial civilization which has the characteristics of technology innovation, emphasizing utilitarianism, nature transformation, wealth accumulation, and market integration, rapidly spread, and dominate the world. Traditional agricultural civilization was proved to be backward and lost in

J. Pan, *China's Environmental Governing and Ecological Civilization*,
China Insights, DOI 10.1007/978-3-662-47429-7_2

the competition with industrial civilization. However, during the understanding and acceptance of industrialization, people have also constantly faced various confusions and rethink: is industrial civilization the ideal social civilization form for human society? Can the fundamental conflicts and problems of industrial civilization be solved?

As early as in the mid of the nineteenth century, in the United Kingdom, the leader of industrial revolution, the economist and philosopher J. S. Mill, started to question the industrial economy which was based on constant extensive expansion. Mill believed that technology and capital could change nature, conquer nature, and produce and accumulate ever-increasing artificial material wealth. However, he also realized that the production and accumulation of material wealth needed to use and consume natural resources; meanwhile, these natural resources were also part of human life, had inherent value, and needed to be reserved. Therefore, he formulated the concept of "static economy" and considered it an ideal economic form.[1] In the "static economy," population size generally remains stable, the total volume and size of economy is more or less stable, and the natural environment also maintains stable. The static economy sought by Mill is not due to resource constraints because he believed that the resource limit is of "indefinite distance" (definite distance vol. 1, p. 220). Mill did not think it reasonable to consider population growth, capital accumulation, and constant improvement in living standard an indefinite process. This was because land was not only for production, it was also the carrier of living space and beautiful nature. In his opinion, sufficient living space and natural landscape were very important. This is because "the serenity and vastness of beautiful nature is the cradle of thinking and belief and it not only benefits everyone, but also is indispensable to the entire society."

In the 1960s, economic development under the direction of industrial civilization caused resource depletion and environmental pollution, forcing people to think about the limit of industrialization and economic growth. In 1962, the American marine biologist Rachel Louise Carson published the book *"Silent Spring"* which described that the large quantity and widespread use of pesticide, a chemical product the industrial society was proud of, could lead to a world without birds, bees, and butterflies. Ms. Carson insisted that the balance of nature was a main force supporting human survival. However, in industrial civilization, chemists, biologists, and scientists believe that human beings have full control of nature. Meadows et al. thought that due to the rigid constraint of natural resources, economic growth had reached its limit; therefore, "zero growth" or even "degrowth" (i.e., negative growth) was needed.[2] The British economist David Preece studied the values of natural resources and believed that apart from use value, natural resources also had choice value and existence value. If natural resources were depleted and damaged, human beings would lose the choice value and existence value of the natural resources whose use value had not been recognized by the market. Mill's "static economy"

[1] Mill, J. S. (1848). *Principles of political economy*, Chapter 6, Book IV.

[2] Meadows, D. H., Meadows, D. L., Randers, J., & Behrens III, W. W. (1972). *The limits to growth*. New York: Universe Books.

was mainly a philosophy thinking. Herman Daly,[3] a criticizer of neoclassic economics, established the theory of "steady-state economics" and argued theoretically that the "steady-state economy" was aimed at keeping population, energy consumption, and material consumption steady or fluctuate in a small range; in other words, the birth rate and death rate of population should be equal and the saving/investment should be equal to asset depreciation. The American economist Kenneth Boulding explicitly confirmed the boundary constraints of the Earth and stated that in the vast universe, the Earth was just a "spaceship" human living in and that the world economy could be nothing other than the "spaceship economy" in a limited space.[4]

Natural scientists often view civilization change from resource constraints and influences. The American electrical engineer Richard C. Duncan[5] calculated per capita demand for fossil fuel and concluded that the life expectancy of industrial civilization was only 100 years. Duncan believed that the current human society was an electromagnetic civilization, and the industrial civilization would collapse in the absence of electricity supply. As electricity generation depends on fossil fuel extraction and combustion, per capita fossil fuel consumption, especially per capita petroleum consumption, has reached its peak, and by 2030, the per capita fossil fuel supply would not be able to meet demand, leading to the end of industrial civilization. Of course, he thought that if other energy sources could replace fossil fuel and human population could stabilize, human civilization could continue. The natural scientist Jared Mason Diamond[6] analyzed the rise and declines of various civilizations in human history, and on the basis on examining 43 civilizations that had disappeared, he concluded five major causes behind civilization collapse, including environmental damage (such as deforestation and soil and water erosion), climate change, dependence on long-distance trade for the supply of indispensable resources, intensification of internal and external conflicts (wars and invasions) due to resource contest, and social reaction to environmental problems. If history development has its own natural causes, industrial civilization will make the above problems occur and overlap in a short period.

The English philosopher Bertrand Russell criticized industrial civilization not based on resource constraint; he thought industrial civilization was against human nature.[7] He believed that the main conflict in industrial societies was not the conflict between socialism and capitalism but the "conflict between industrial civilisation and human nature." He thought industrialism caused too much waste of world

---

[3] Daly, H. (1991). *Steady-state economics*, 2nd Edn. Washington, DC: Island Press, p. 17.

[4] Boulding, K. (1966). The economics of the coming spaceship Earth. In H. Jarrett (Ed.), *Environmental quality in a growing economy* (pp. 3–14). Baltimore: Resources for the Future/ Johns Hopkins University Press.

[5] Duncan, R. C. (2005). The Olduvai theory: Energy, population, and industrial civilization. *The Social Contract*, Winter 2005–2006: 1–12.

[6] *Collapse: How societies choose to fail or succeed.* New York: Penguin Books. 2005: ISBN 0-14- 303655–6.

[7] *The prospects of industrial civilization* (in collaboration with Dora Russell). London: George Allen & Unwin. 1923.

resources, and the ultimate hope for human beings should be scientific outlook. Russell's thinking about industrial civilization's damage to human nature was inherited by the ecological Marxists or ecological socialists in the 1970s.[8] They pointed out that the belief of "controlling nature" established under the capitalist system was the root cause of ecological crisis and believed that consumption alienation would occur in capitalist societies and proposed to build a "society easy for survival" to solve the ecological crisis. Since then, there have also appeared the theory of economic and ecological dual crisis, political ecology theory, economic reconstruction theory, and ecological socialism theory, forming a systematic ecological Marxism theory. Although the detailed ideas of ecological Marxists are different, they all tried to reveal the origins of ecological crisis from the perspective of Marxism theory and to find a way out for human beings from the ecological difficulties. The key question they focus on was the relationship between human society and nature, and they believe that capitalist system and production pattern are the origin of ecological crisis. They criticize technology rationality and consumption alienation, and their ideal is to build a society of human freedom and harmony between human and nature. They believe that the economic rationality under industrial civilization will inevitably make the labors lose their human nature and become machines, inter-person relations monetary relation, and the relationship between human and nature a tool relationship. Ecological Marxists' criticism absorbs the ecological theories of environmentalism, ecological ethics, and postmodernism. From ideology perspective, ecological Marxists attributed the origin of ecological crisis to capitalist system itself. They try to lead the ecological movement with Marxism and find a new way out for socialism. Ecological socialists believe that ecological problems are in fact social problems and political problems; ecological problem can only be fundamentally solved through abolition of capitalism. They are dedicated to the combination of ecological principles and socialism and try to transcend capitalism and traditional socialism and build a new socialism form for harmony between human and nature. Ecological socialists regard the fundamental conflict of capitalist society is the "fundamental conflict between capitalist production and the entire ecosystems" and believe that ecosystem deterioration is the inherent logics of capitalism. Therefore, the only solution to the problem is smashing the logics itself. Globalization accelerates the transfer and spreading of ecological crisis; environmental problems have already crossed national and regional borders and become a common problem of the entire mankind. Therefore, to solve the problem, common understanding is necessary; to establish common understanding, justice is needed; to realize justice, the existing inequitable international order controlled by advanced capitalist countries has to be changed; to change the existing international order, socialism has to be promoted as the essence of socialism is justice.

---

[8]The representative scholar William Leiss, a Canadian scholar, who published *The Domination of Nature* in 1972 (Leisse, W. (1972). *The domination of nature*. New York: George Brazillerinc; 1994, McGill-Queen University Press) and the *Limit to Satisfaction* in 1976 (Leiss, W. (1976). *The limit to satisfaction*. The University of Toronto Press; 1988, McGill-Queen University Press).

The above analysis indicates that the western scholars have questioned and criticized the rationality of industrial civilization from natural, environment, philosophy, and ideology aspects. They have in-depth understanding about the disadvantages of industrialization. Some scholars have also proposed some measures and solutions, for example, human beings should consciously choose to live in harmony with nature, countries should realize the zero growth "static or steady-state economy," and match the capitalist systems in line with industrial civilization. They have realized that industrial civilization was just a phase in human civilization development, and it would inevitably come to an end due to its inherent properties. One scholar, Paul Bohannan clearly pointed out in his book *"Transcend Civilisation"* published in 1971 that "obviously we are on the threshold of post-civilisation. When we solve the current problems, the society and civilisation we will build will be unprecedented. It may be, to a big or small extent, more civilised than what we have now, but it will definitely be different from the civilisation we know."[9] Although he foresaw that a brand new civilization form would appear, he did not say how the civilization would look like. Ecological Marxists also mentioned social civilization reform and thought the future society should be a revolutionary one in human civilization history, and it should be a new society combining economic efficiency, social justice, and ecological harmony. Yet the various criticisms mainly focus on solving certain problems and addressing the problem from one aspect. They did not comprehensively analyze the situation from development paradigm perspective. Therefore, it is impossible to comprehensively implement the above proposed solutions under the industrial civilization paradigm.

2.2 The Origin of Ecological Civilization

Ecology refers to the interrelations and existence of individual life forms, species, community, as well as life and environment. It is natural living status and a natural phenomenon. Civilization is a creation by human society. As early as prior to 100 BC, Yi-Ching, the Chinese book of changes, had pointed out the real origin of civilization is the human body. In the Qian Hexagram of Yi-Ching, it was said "seeing dragon in the field, then the whole world is bright." It means that a person's spiritual dragon can be seen in the public region of the human body. The appearance of the dragon can constantly increase the brightness of human body and lighten up the inner body. It can also transcend the body and light up and influence people around one. The human spirit and the brightness enhance each other and bright up a person's mind.[10] In the Tang Dynasty, Yingda Kong commented on "Shangshu," the Chinese book of history, and further explained "civilisation" as "ordering sky and

[9] Bohannan, P. (1971, February). The state of the species, beyond civilization. *Natural History* – Special Supplement 80.

[10] Xiong, C. J. (2010). *"Civilisation and education of the dragon culture."* Unity Press. ISBN 9787512601901.

earth and lighting up the four directions." "Ordering sky and earth" is changing nature and belongs to material civilization; "Lighting up the four directions" means dispels stupidity and belongs to spiritual civilization. Obviously, in ancient China, people's definition of civilization is different from culture. Culture is mainly improvement of a person's internal attainment, while civilization includes both material and spiritual contents and is a combination of both material and spirit.

In the western language system, the literal meaning of "civilization" originated from *civilis* which is the root of the word *civil* in Latin. *Civis* is citizens, while *civitas* refers to cities. When people come together, there needs to be certain behavior rules and therefore forms civilization. No matter in the East or in the West, the origin and carrier of civilization is human; without human, there is no civilization. Civilisation also means enlightened and is opposite to ignorance. However, in the Chinese language, civilization more emphasizes a person's internal cultivation and capability improvement; while in the European language, it more stresses people's external appearance, inter-person relations, or social groups.

However, the creation of civilization is the results of human and nature interactions, and no civilization can transcend nature. Some civilization with distinctive territory characteristics, such as the Maya civilization, the ancient Egyptian civilization, and the 5000 years of continuous Chinese civilization, is the natural, material, and spiritual aggregation of people living in a certain geographic environment and the local natural conditions they fit in. Civilisation evolution, which features specific productivity and production patterns, often has technology annotations. For instance, the primitive civilization was of low productivity and based on the production pattern of hunting and gathering, while the farming civilization is based on independent production using natural resources.

In the mid of the twentieth century, with the occurrence of a series of severe pollution accidents, people started to think about the shortcomings of industrialization and pay attention to environmental protection and sustainable development. From the publishing of *the Silent Spring* in 1962 to the release of Limit of Growth and the United Nations "Conferences on the Human Environment" organized in Stockholm in 1972 and then to the United Nations "Environment and Development Conference" in 1992 and the "World Summit on Sustainable Development" in 2002, the international communication has been searching for a new industrialization approach and the path for sustainable development that combines and coordinates economic development, social progress, and environmental protection. Although environmental protection and sustainable development have been included in national and international political agenda, these efforts fail to rethink development paradigm from the perspective of social civilization forms. Existing literatures indicate that Roy Morrison was the first scholar who regarded ecological civilization as the civilization form that would replace industrial civilization. Morrison published his idea in a book in 1995,[11] but his main argument is from the perspective of democracy.

[11] Morrison, R. (1995). *Ecological democracy.* Boston: South End Press.

In China, the concept of ecological civilization was first formulated by Qianji Ye, an agricultural economist, in 1984.[12] Mr. Ye defined ecological civilization from the perspectives of ecology and ecological philosophy. The scientific dimension of natural ecology originated from the West, while the philosophy dimension of human ecology has thousands of years of history and development in China. As early as over 2500 years ago, in Tao TeChing,[13] Lao Tze, the founder of Taoism, formulated the philosophy that "Man takes his law from the Earth; the Earth takes its law from Heaven; Heaven takes its law from the Tao. The law of the Tao is its being what it is." It means that "people live, work and reproduce according to the law of the earth; the earth changes seasons and produce various life according to the laws of the heaven; the heaven operates and changes according to the major 'Tao'; while the major 'Tao' follows the law of nature and lets nature takes its course."[14] Obviously, Lao Tze was trying to explain the relations between human and nature and the rules governing them. Here the key is the meaning of the word "law." One of its meanings is "following the example." However, its other annotations include "depending on," "being up to," and "complying." Human beings' survival and development relies on and is up to the produce of the Earth, while mother earth's produce depends on and is up to the heaven, i.e., the change of four seasons and the weather, while functioning of the "heaven" depends on and follows the "Tao," which means natural laws and rules; "Tao" needs to follow the laws of nature or "what it is" or naturally. In summary, the relations between human and nature is not the human transforming and controlling nature under the industrial civilization but human respecting nature and complying with the laws of nature. Later, Zhuangzi further developed the ideas of Lao Tze into the philosophy thinking system of "harmony between human and nature." Xianlin Ji interpreted this thinking system as the mutual understanding and friendship between nature and human beings and thought of this as the Chinese culture's most important contribution to human civilization.[15] This thinking is a distinct contrast to the western philosophy of using advanced science and technology to conquer, loot, and pillage nature.

For over 2000 years, the Chinese farming civilization followed the philosophy of "harmony between human and nature," complied with the laws of nature, left space for nature to restore itself in economic development, and created several periods of glory in world development history. Despite natural disasters and human mistakes, the Chinese civilization developed continuously and contributed to world civilization development. After the birth of industrial civilization, industrial countries used advanced science and technologies to assault and pressed the Chinese traditional farming civilization. The Chinese practice of pursuing harmony between human

[12] Ye, Q. (1984). *Ways of training individual ecological civilization under nature social conditions.* Moscow: Scientific Communism, 2nd issue.

[13] Tze, L. (600 BC – 470 BC). Whose family name is Li and given name is Er and who styled himself as "Boyang." Tao Ching Chapter 25 in *Tao TeChing*.

[14] This interpretation is from the China Culture Web, Sinology Class "Sinology Classics – One Sentence per Day." www.wenming.cn 2010-01-07.

[15] See "Harmony between human and nature" at: http://baike.baidu.com/view/4259.htm?fr=aladdin.

and nature could not resist industrial development which has enormous nature-transforming capability. China's Great Leaping Forward in the late 1950s and Campaign of Following the Example of Dazhai in Agricultural Development in the late 1960s were efforts to imitating industrialization and reflected the belief that human being can always win the competition with nature. The results of these imitating attempts were damages to natural environment, soil and water erosion, resource degradation, and ecosystem imbalances. The discussions on ecological civilization and ecological economics study started in the late 1980s in China were in fact targeting at addressing ecosystem degradation, neither environmental pollution nor the depletion of resources, especially fossil fuel. At that time as China did not have the high energy consumption and high pollution from industrialization, the ecological degradation was reversible.

Since entry into the twenty-first century, China's rapid and large-scale industrialization has exceeded the carrying capacity of its resources and natural environment. China's total consumption of commercial fuel was only 650 million tonnes of coal equivalent in 1980; this figure had reached 3.75 billion tonnes of coal equivalent by 2013. In 1992, China was still a net exporter of crude oil, but in 2013 its petroleum import had exceeded 320 million tonnes and its coal import was 330 million tonnes. Vast areas of China are covered in smog, its lakes and rivers are polluted, its soil is poisoned with heavy metals, the groundwater in North China is over extracted, and large amount of high-quality arable land in South China has been occupied for irreversible industrial development and urban construction. Before the reform and opening up in 1978, China's agricultural production profile is grain production in the south supporting the north; now it has changed into rice from north (east) feeding the south.

The industrialization copying the western production patterns leads to rapid accumulation of material wealth, but ecosystem imbalance also reaches the tipping point of abrupt changes. Ecological civilization has again been included in the governmental agenda as a key task. In 2002, at the 17th Congress of the Chinese Communist Party, the task of building ecological civilization was decided which boosted the academic community's widespread and in-depth studies on ecological civilization.

2.3 The Connotations of Ecological Civilisation

Literally, ecological civilization is the combination of ecology, a natural concept, and civilization, a human concept, and the core of ecological civilization is the relationships between human and nature. The origin of ecological civilization that was attributed to the Chinese philosophy thinking of "harmony between human and nature" is because of the necessity to understand and define the relationships between human and nature.

The direct and most fundamental reflection of relationships between human and nature is shared values: how people see and treat nature. The thinking of "harmony between nature and human being" by ancient Chinese thinkers emphasizes the integrity of human and nature, and human being as part of the mother Earth must respect and comply with nature. Human is not the master of nature and should not transform and conquer nature. Instead, the relationship between human and nature is equal and harmonious; therefore, human being should treat nature rationally and justly. What human should pursue is not the indefinite accumulation of nonnatural material wealth but accepting and respecting the value of nature. If respecting and complying with nature is some kind of ecological justice, the value of harmony between human and nature also includes social justice, in other words, respecting human rights and justly sharing natural resource benefits. Ecological justice and social justice complement each other and form the value foundation of ecological civilization. The common narrow understanding of ecological civilization from cultural dimension is limited to shared values. It emphasizes that human is part of nature, and in principle human should respect nature and treat nature justly. In behavior rules, all human activities should fully respect natural laws, and human should seek harmony development with nature.

In broad sense, ecological civilization not only includes the value of human respecting nature and coexisting and co-prospering with nature but also covers the production patterns, economic foundation, and governance systems based on the shared value. It is a social civilization form covering all material and spiritual achievements of the society of "harmony advancement between human and nature, highly advanced productivity, all-around human development, and sustained social prosperity."

Its production pattern is not a linear pattern, that is "from resource processing to products, from products to wastes". Instead, it is based on the precondition of ecological rationality and seeks efficiency of material output instead of the simple benefit of output maximization. This requests the society to get rid of the production pattern that is based on low efficiency, extensive development, and looting and damaging nature and to produce in the way of high resource utilization efficiency and minimum environmental damage, so as to realize the production pattern from raw material to products from raw material processing to raw materials. In ecological civilization, the consumption pattern does not pursue material occupation and a luxury and wasteful life but a green, economic, healthy, rational, and quality life. Human's basic needs for materials are limited, while human desire can be indefinite. The lifestyle under ecological civilization does not request people to be stingy with daily necessities and coming back to the primitive life in the age of farming civilization. Instead, it is on the basis of securing the satisfaction of people's basic material needs, restraining the desire of unnecessary material occupation and consumption. The demand determined by social members' lifestyle influences production patterns. Therefore, the life and consumption styles in ecological civilization are the specific reflection and embodiment of the ethics and shared values of ecological civilization.

In ecological civilization, the ultimate objective is comprehensive development and security of quality life for the people. The ecological civilization in modern age is no longer the passive accommodation of nature over 2000 years ago but the harmony between human and nature based on modern science and technology and people's deep understanding about nature. The beauty and vastness of nature is not only the cradle of thinking but also a necessary element for quality life. Beautiful nature does not require economy shrinkage and social recession. Instead, maintaining the beauty of nature also needs economic prosperity and social stability. The harmony between human and nature on the basis of modern science and technology and economic development level is an ideal situation that combines the harmony between human and nature, among people, and between individuals and society.

In social system, there is a set of effective mechanisms and legal systems for respecting and protecting nature, promoting social justice, regulating production and life patterns, and securing human development. People's shared values need to be regulated and guided by social systems. The formation and development of production and life patterns in ecological civilization must also be subject to the regulation and guidance of systematic social mechanism and legal systems. Human development and life quality also needs to be regulated by relevant standards, rules, and legal systems. Capitalist institutional and legal systems are both the product of industrialization process and the safeguard for the smooth progressing of industrialization. Without market systems and legal systems, capitalism development would not have been so efficient and systematic. The above discussions indicate that as a development paradigm, ecological civilization has the core elements of justice, high efficiency, harmony, and human development. Justice is to realize the ecological justice of respecting the rights and interests of nature and the social justice of protecting human rights and interest. High efficiency is to pursue the self-balance and ecological productivity efficiency of natural ecosystems; economic production systems' economic efficiency of low input, no pollution, and high output; and human social systems' social efficiency of comprehensive legal systems and norms and smooth functioning. Harmony is to achieve the compatibility and mutual interests between human and nature, among different people, and between individuals and society. It includes the balance and coordination between production and consumption, economy and society, and rural areas and urban areas as well as among different regions. The main parameters of human development include dignity, quality, and health of life. All the elements are interrelated: justice is the necessary foundation of ecological civilization, efficiency is the means for realizing ecological civilization, and harmony is the exterior appearance of ecological civilization, while human development is the ultimate target of ecological civilization.

2.4 The Orientation of Ecological Civilization

China's understanding about ecological civilization has also experienced a process of deepening and clarification. First of all, what are the relationships between ecological civilization and material civilization, spiritual civilization, and political civilization? In the modern civilization of socialism with Chinese characteristics, ecological civilization is a new term appeared after material civilization, spiritual civilization, and political civilization. The 17th National Congress of the Chinese Communist Party clearly decided "to build ecological civilisation and generally establish the production structure, growth approach, and consumption pattern of energy resource saving and ecological environment protection" and to boost the solid establishment of ecological civilization perception. Ecological civilization was attached with the same importance as material, spiritual, and political civilization.[16] In the past, people's understanding about ecological civilization is concrete measures and often limits to nature protection, resource conservation, and pollution control. Just as Hu Jintao, then Chinese president, stated after the 17th National Congress of the Chinese Communist Party, "building ecological civilisation, in essence, is building a resource efficient and environmental friendly society that is based on resource and environment carrying capacity, complies with natural laws, and aims as sustainable development."

In this way, ecological civilization, together with material civilization, social culture, spiritual civilization, and political civilization, forms the "five dimensions" of socialist society.[17] Economic construction, political construction, cultural construction, social construction, and ecological civilization construction supplement each other, and none of them should be neglected. Good ecological environment is the important foundation for the social development of social productivity and continuous improvement of people's life quality. Building ecological civilization also needs the important conditions of material civilization, spiritual civilization, and political civilization. More importantly, economic, political, social, and cultural building are all related to the contents of ecological civilization building, and it is necessary to integrate ecological civilization in the various aspects and whole process of economy, politics, society, and culture. Some people believe that ecological civilization is promoting traditional Chinese culture. Many scholars tried to study the cultural origin of ecological civilization in Confucianism, Buddhism, and Taoism and believed that Confucianism, Buddhism, and Taoism explained the relations between human and nature from different perspectives.[18] The ecological ethic thinking of "harmony between human and nature" contains such shared values as respect to nature, universal love to everything, as well as harmonious relations between human and nature. It includes rich ecological civilization thinking and is important reference for building modern ecological civilization system.

As a development paradigm, ecological civilization needs to be understood in broad sense. It is not only a kind of shared values but also a social civilization form that is based on the shared values and embodies it in production relationship and

---

[16] Yu, M. (2006). Ecological civilisation – the fourth civilisation of human society. *Green Leaves*. [J], (11); Jianjun Zhao. Building ecological civilisation is the requirements of modern age. [N]. *Guangming Daily*, 2007-08-07.

[17] Hu, J. (2012). Report to the 18th National Congress of the Chinese Communist Party.

[18] Junhua Ren. (2008). The ecological ethic thinking of Confucianism, Taoism, and Buddhism [J]. *Human Social Sciences,* (6); Min Zhang. (2008). Ecological civilisation and its contemporary value. PhD Thesis of the Graduate School of Chinese Communist Party Central School; Yu Kang. (2009). A comparison on the thinking of Confucianism, Buddhism, and Taoism [J]. *Tianjin Social Sciences,* (2).

institutional and legal systems. It inevitably contains the compatible contents of material civilization, spiritual civilization, and political civilization. Ecological civilization includes the contents of material civilization such as ecological economy, green recycle technologies, tools, instruments, and ecological welfare. The contents of spiritual civilization include ecological justice, ecological obligations, ecological awareness, laws, institutions, and policies as well as ecological democracy and other contents of political civilization. In this view, ecological civilization is an overarching and integrated concept. This means that ecological civilization is a new development stage of human society, and if measured with technology parameters, it is a new stage after primitive civilization, agricultural civilization, and industrial civilization.[19] In temporal view, primitive civilization existed around one million years during the Stone Age, when people had to rely on collective power for survival, and the material production activities were mainly simple gathering, fishing, and hunting. During the period of agricultural civilization, the discovery and creation of iron tools led to dramatic improvement in people's capability of changing nature and it lasted around 10,000 years. Industrial civilization is the modern human society ushered in by the industrial revolution starting in the United Kingdom during the eighteenth century, and it has lasted 300 years. Its excessive resource consumption and huge damages to environment mean that this social civilization form cannot last as long as primitive civilization and agricultural civilization. A new social development paradigm or civilization form is urgently needed. However, the social form of ecological civilization is different from the social civilization forms of slavery civilization, feudal (medieval) civilization, capitalist civilization, and socialist civilization as it emphasizes the common future of human being.

What are the relationships between green economic, recycle economy, low-carbon ecology, and ecological civilization? Conceptually, green economy is a production and consumption economy systems that takes green development as orientation and measurement and bears the characteristics of protecting human living environment, reducing and getting rid of pollution, protecting ecosystems, and promoting sustainable energy use and human health. Recycle economy refers to the big system of human and nature, changing the traditional economy of linear increases in resource consumption to an economy system of closed recycle through reducing, recycling, and reusing in the whole process of resource exploitation and mining, product processing, consumption, and disposal. Low-carbon economy is an economy form with high-carbon productivity and human development that is aimed at controlling greenhouse gas emissions and realizing the global vision of limiting global ground temperature increase to maximum 2 above the preindustrial revolution level.

[19] Many scholars share this view. Among them, the representative ones include: Keping Yu. (2005). Scientific development views and ecological civilisation [J]. *Marxism and Reality,* (4) Weihong Li. (2006). Ecological civilisation—the ineivtable path of human civilisation development [J]. *Socialism Study*, (6), 114—117; Zhihe Wang. (2007). Chinese emphasis on harmony and design of post-modern ecological civilisation [J]. *Marxism and Reality*, (6); Ruiqing Chen. (2007). Building socialist ecological civilisation to achieve sustainable development [J]. Inner Mongolia United Front Theory Study, (2).

Ecological economy is an economy system that, on the basis of following ecological principles, is both satisfying human society's development demand and maintaining ecosystem balances through ecological designing and engineering technologies. Therefore, all the above concepts from different aspects emphasize improving resource and energy utilization efficiency, reducing pollution emissions and discharges, and realizing sustainable development. As a human civilization form, ecological civilization stresses harmony development between human and nature on the whole. Green economy, recycle economy, low-carbon economy, and ecological economy are all approaches and means for building ecological civilization from a certain aspect.[20]

As a civilization form, ecological civilization has such prominent features as overarching, integration, multiple dimension, and sustainability. Each social civilization form is the aggregation of material and spiritual achievements of certain social development stage and are primary and as a whole. Ecological civilization, which will replace industrial civilization, must be supported by suitable material, spiritual, and political civilization. Other contents of social civilization construction, such as material civilization, are not primary but secondary to the framework of ecological civilization. Ecological civilization needs to be based on certain material civilization and to guide and regulate material civilization. Moreover, every civilization form is comprehensive and covers many contents, and these contents are not sporadic and isolated but interrelated and integral. However, the integrity of ecological civilization covers more aspects of natural ecosystems, and it puts ecosystems in parallel and equal position with economic and social systems, while industrial civilization regards natural systems only as the object for transformation.

Like other civilization forms, ecological civilization contains multiple dimensions and is inclusive. Ecological civilization is not simple negation of agricultural civilization and industrial civilization. Instead, it improves and upgrades them and absorbs the scientific contents of agricultural and industrial civilizations. The most prominent feature of ecological civilization is sustainability. Agricultural and industrial civilizations are unable to address the problem of unsustainability. They either neglect or are helpless in addressing ecosystem degradation and environmental pollution. Ecological civilization emphasizes harmony and respect to nature and green development. It not only protects the sustainability of natural ecosystems but also demands sustainability in economic and social development.

---

[20] Shunji Huang, Zongchao Liu. (1994). Ecological civilisation views and China's sustainable development [J]. *Science and Technology International*, (9); Kai Zhang. (2003). Recycling economy as the inevitable path toward ecological civilisation [J]. *Environmental Protection*, (5); Xiangrong Liu. (2009). Ecological perspective on economic development patterns and China's ecological civilisation building [J]. *Nanjing Social Sciences*, (6); Chongzheng Xu, & Weiran Jiao. (2009). Building ecological civilisation, developing recycling economy, and human development [J]. *Reform and Strategy*, (10).

2.5 Negation to Industrial Civilization?

It has been mentioned in the previous section that ecological civilization, as a development paradigm originated from ancient Chinese philosophy and production practice, is multidimensional and inclusive. The shaping, development, and replacement of social civilization forms are a gradual process, and different stages can combine, overlap, and interweave. In other words, during each stage of human history, several civilization forms can often coexist, learn from each other, and compete with each other. One of the civilization forms gradually becomes dominant and forms a development paradigm with its own distinct characteristics. The history of human civilization progress indicates that at each social development stage, there is always one dominant civilization form and some secondary civilization forms.

The inherent limitations and shortcomings of industrial civilization call for the creation of a new civilization form. Does this mean that all elements of industrial civilization have to be denied or replaced? Industrial civilization, which originated from industrial revolution, was born in agricultural civilization. Due to science and technology innovation, fossil fuel use, and major improvements in production tools, industrial civilization helped mankind break away from being afraid and at the mercy of nature in "the realm of necessity" in resource use. Powered by fossil fuel, industrialization process has led to rapid increase in social productivity and creates huge material wealth that is needed for human social development and has fundamentally changed the ideas and values, cultural form, production, and life patterns. Through enhancing and consolidating these changes in governance systems, it has realized major changes in social development paradigms and brought about dramatic changes in economy, politics, culture, living environment, social structure, and people's lifestyles. Agricultural civilization, which has the characteristics of fearing nature, low productivity and urbanization rates, self-dependence, and a thrifty lifestyle, was gradually dropped by societies. In its place came industrial civilization which has the properties based on utilitarian values, led by technology innovation, and the production patterns, lifestyle, and social system structure of conquering nature, obtaining resources and extravagant consumption. Industrial civilization became the leading civilization form of industrial societies.

However, the shared values of utilitarianism and fetishism, limited fossil fuel, and the linear production patterns "from raw materials to waste," the increasing inequality in income distribution, as well as the pathetic consumption patterns of pursuing extravagance and luxury make industrial civilization be plagued by the crises of resource depletion, environmental pollution, ecosystem degradation, and social conflicts. Industrial civilization is unable to continue, and a new development paradigm of sustainable prosperity is fermenting. A brand new development paradigm is needed to understand the limitations and shortcomings of industrial civilization, rethink the shared values of industrial civilization and the production and life patterns shaped by them, and improve the governance systems of industrial civilization. Although some elements of the agricultural civilization, such as being in awe of nature, are positive, it is impossible and unnecessary for human civilization to go back to the farming social forms of "slash-and-burn farming" and extreme material scarcity. Using the shared value and governance system of harmony between human and nature to reform and improve industrial civilization, a new social civilization form or development paradigm will come into being. Such a civilization change means ecological civilization won't simply deny or replace industrial civilization. Instead, it is a kind of "transplantation" and "developing the useful and discarding the useless." Ecological civilization construction needs to inherit and develop the science and technology achievements and management techniques of industrial civilization; absorb some useful system and mechanism elements of stimulating and protecting innovation, discarding the shared values of industrial civilization; and improve the improper production patterns and lifestyles.

To accept and realize the development paradigm of ecological civilization, won't it be necessary to terminate or abolish industrialization? This is a misunderstanding. China's industrialization has made enormous progress. However, its industrialization level is still at the mid and late phases of industrialization and its development level is still low. On one hand, ecological civilization is neither to fetter productivity nor to impede and terminate development. On the other hand, it is difficult for the development paradigm of industrial civilization to continue or sustain, and it must transform. Industrialization liberates productivity and innovation and accumulates material wealth; therefore, it shall not be denied. Instead, it should continue. However, the industrialization that needs to be continued is not the traditional one of high pollution and low efficiency; in contrast, it should be environmental friendly, high quality, and sustainable. Hence it is necessary to use the development paradigm of ecological civilization to transform, upgrade, and reshape industrialization. Because civilization development is a gradual process, the change of development paradigm is also a process from quantitative change to qualitative change. Ecological civilization building does not wait to start until industrialization is completed. On the contrary, it is necessary to shape and enhance the shared values of ecological civilization, improve production patterns and lifestyles, and eliminate the shortcomings of industrial civilization, so that the Chinese economic and social development can avoid falling into the trap of unsustainability.

If industrial civilization and ecological civilization are two different development paradigms, then what are the differences between them? The first difference is the shared values or ethic foundation. During the early days of industrial revolution in the United Kingdom, David Hume, the leading thinker of Scottish Enlightenment,[21] formulated utilitarianism from the ethics and sensibilities in his *Treatise of Human Nature*. Later, Jeremy Bentham, an English thinker,[22] attracted by the strong persuasiveness of utilitarianism, put forward the principle of "the majority people's maximum happiness," also known as the "principle of maximum happiness." John Stuart

[21] Hume, D. 1711–1776, is a Scottish philosopher, economist, and historian. He was considered one of the most important scholars in the Scottish Enlightenment Movement and western philosophy history.

[22] Bentham, J. (1748–1832). English jurisprudent, utilitarian philosopher, economist, and social reformist.

Mill, a philosopher and economist, in the book *Utilitarianism* he published in his old age, stated that human being is able to sacrifice his own maximum happiness for other people's happiness; if sacrifices cannot increase total happiness or are not inclined to increase total happiness, they are in vain. He emphasized that in the behaviors of utilitarian, the standard happiness is not the actor's own happiness but the happiness of all related people. When one treats others in the way one expects others to treat oneself and one loves one's neighbors like loving oneself, utilitarian ethics reaches its ideal perfection state. Obviously, happiness is based on materials and realistic; if some components of environment or resources cannot bring happiness, they do not have utility. Moreover, utility is realized through market. If something has no market value or low market value, it has to be replaced by things with utility or higher utility. The ethical foundations of ecological civilization inherited the ancient Chinese philosophic thinking and sought ecological justice and social justice. Human being is part of nature and should not damage nature for the sake of its own happiness. Although utilitarianism also stresses maximum happiness of individual collections—societies—it neglects justice among people and between individuals and societies. The thinking of some scientists, for example, the "*Origin of Species*" of Charles Darwin, based on analysis and inductions from the natural science perspectives, highlighted natural selection and survival of the fittest. The social Darwinism, which was based on the Darwinism in natural science, not only ignored the interests of weak social groups but also was used by racists to promote racially biased policies.

In terms of objective selection, the development paradigm of industrial civilization pursues profit maximization, wealth accumulation maximization, and utilization maximization, leading to whole societies' GDP worship and competition for interests. As a development paradigm, ecological civilization is aimed at harmony between human and nature, environmental sustainability, and social prosperity, and it ignores the accumulation of monetary returns and material assets. In fact, ecological civilization attaches greater importance to natural assets which do not need maintenance investment and can naturally maintain or increase their values; in contract, artificial assets need huge amount of maintenance costs and constantly depreciate and lose their values.

In view of energy foundation, industrial civilization relies on fossil fuel, while ecological civilization emphasizes sustainability and seeks for the transition toward sustainable energy. Of course, under the development paradigm of ecological civilization, energy production and consumption won't go back to the low efficiency and low-quality utilization of renewable energy under agricultural civilization; it will be high efficiency and high-quality services of commercial energy. Industrial civilization denies or ignores the boundary of development and believes in unlimited extensive expansion and the absence of rigid constraints of resources. Ecological civilization clearly admits the existence of natural limit and complies with the rigid constraints.

In production and consumption patterns, the extensive linear mode of resource—production process—products and wastes under industrial civilization will be replaced by the recycling mode of high-efficiency raw material—production process—products and raw materials under ecological civilization. Under industrial civilization, consumption is occupation and wasteful luxury consumption, while under ecological civilization, it is low-carbon, quality, healthy, and rational consumption.

These differences between ecological civilization and industrialization also indicate the elements of industrial civilization that need to be transformed and improved. While the many advantages of industrial civilization will not only be inherited but also be further developed under ecological civilization, the technology innovation and support under industrial civilization will undoubtedly be needed under ecological civilization. However, the development paradigm of ecological civilization needs to differentiate technologies. Technologies that can contribute to sustainability and efficiency improvement should be encouraged, while technologies that lead to damages to nature and wasting of resources, such as the technologies for building height competition, need to be restricted or forbidden. Industrial civilization's governance and legal systems based on the utilitarian principles, for instance, democracy, the principle of law, and market mechanisms, can be directly inherited or transplanted by the development paradigm of ecological civilization. However, ecological civilization paradigm also has its unique governance system contents such as ecological compensation, ecological red lines, as well as appraisal and evaluation of natural resource inventory and changes.

2.6 Ecological Civilization Construction

In the course of industrial civilization reflection and criticism, the practice of green development and environmental protection has been making progress. At the level of national politics, boosted by the tides of green and ecological thinking, ecological movements have been taking place in some countries. In May 1972, the first national green party in the world—Value Party—was established in the New Zealand. In 1980, a "Green Party" with clear political creeds was founded in Germany and chose "ecological socialism" as its political target. After that, green Parties appeared and continued to expand, and at its peak they had seats in the parliaments of over 20 countries and some of them even came into power. During the 1990s, ecological socialism created the new concepts of "red parties" and "green parties"; put forward the political, economic, social, and ideological opinions; and formed its thinking system and political manifestos.

Because of the importance of environment and sustainability for mankind's common future, plus the close interrelations between environment and development, sustainable development is becoming an increasingly important issue on international political agenda. In 1972, the United Nations held its first conference on environment in Stockholm in order to find solutions for the pollution and damages of human settlement environment. This was the first international alarm about environmental problems, and it also promoted debates and criticisms about the development paradigm of industrial civilization. The report of "Our Common

Future" submitted by the United Nations Environment and Development Commission in 1982 for the first time defined the three pillars of sustainable development: social justice, economic efficiency, and environmental sustainability. In 1992, the United Nations organized its Conference on Environment and Development in Rio de Janeiro in Brazil. The conference passed the *21st Century Agenda* for promoting sustainable development worldwide, signed the international conventions on climate change and biodiversity protection, and created international institutions and laws on global environmental protection.

In 2000, the United Nations formulated the Millennium Development Goals in order to eradicate poverty and secure natural resource supply for social and economic development. In 2012, the United Nations Sustainable Development Summit was convened in Rio de Janeiro of Brazil. Green transition became the core subject of this summit, and the global community not only announced war against poverty but also clearly requested countries to establish sustainable development targets, include sustainable production and consumption, and establish institutional and legal arrangement in their development objectives. These international developments have boosted the environmental protection in China, yet China's ecological civilization construction is in response to the urgent needs of domestic development transition and a responsible contribution to global ecosystem security.

China's industrialization started later than many other countries. After the industrial civilization, China tried to learn from the West and imported some technologies to develop industries. Yet in those days, China was in the stages of semicolonial and semifeudal society, new democratic society, and the early phase of socialist society; its traditional culture was impacted and shocked by western industrial civilization, and China's industrialization was to a large extent a passive response to external pressure. After the foundation of the People's Republic of China in 1949, China was still a backward agricultural economy, and providing food and clothing for its large population was the new Republic's top priority, and the damages to nature were still traditional and reversible. Consequently, environmental protection and ecological civilization construction were not included as an important issue on the government's agenda.

After China's reform and opening up in 1978, Deng Xiaoping set the strategies of three-step development. By 2000, China had already realized objectives of the first two steps. China's GDP was 5.5 times the level in 1980 and exceeded the "quadrupling" target, and people's living standard reached relative well off. The target of the third step is to make people's living standard to be relatively high, generally realize "four modernizations," per capita GDP reaching the levels of medium developed countries by the mid of the twenty-first century.

According to the classification criteria of the World Bank, China's per capita GNI exceeded 6000 USD in 2013 and approached the level of upper middle countries. However, realization of the targets for the third step is not just national average per capita income; it also demands relative equality in social income distribution. In 2013, the per capita disposable income of urban resident is 3.13 times the level of rural residents. Major income gaps between rural and urban residents, the rich and the poor, and different regions of China call for narrowing income gaps and realizing common prosperity.

In terms of industrialization and urbanization progress, China's economic and social development has now reached a critical point for social and economic transition. Per capita income and economic structure data indicated that in 2010, Beijing, Shanghai, and Tianjin had entered the stage of post-industrialization; some provinces in the Yangtze Delta, the Pearl River Delta, and Bohai Sea Rim are in the late stage of industrialization. In 2013, China's steel production reached 883 million tonnes and its cement production 2.09 billion tonnes; China's energy consumption became the biggest in the world, 20 % more than that of the United States and 50 % more than the total energy consumption of the 27 members of the European Union. Fifty-five percent of China's crude oil depended on import, and its iron ore import was as high as 686 million tonnes. These figures show that the space for further quantity expansion of China's raw material industry and manufacturing sector will be very limited due to resource and market constraints. China has to change its production and consumption patterns. Industrialization has boosted China's economic development and modernization. However, industrial civilization, which has the characteristics of focusing on profit maximization, damaging environment, neglecting equity, and wasting resources, obviously is not the shared value of socialism.

In 2013, China's urban population had reached 720 million, accounting for 53.73 % of the national total. This means that China has entered the age of urban society on the whole. China's urbanization will continue, and it is expected that by 2030, it will reach around 70 %, and China's urban population will be more than one billion. If China follows the urban production and consumption patterns of developed countries, China's resources and environment won't be able to support the simple size expansion of cities. As a responsible big country, China needs to not only secure its sustainable development but also contribute to global sustainable development.

In view of China's current industrialization stage, urbanization level, and the needs of global sustainable development, Chinese economy and society are facing the severe challenges of "imbalance, inharmony, and unsustainability," and these factors will significantly impede the realization of China's third-step development targets. To build a society of socialism with Chinese characteristics, China should not and cannot copy the capitalist "industrial civilization." Instead, China needs to, on the basis of absorbing the scientific and rational contents of industrialization, use ecological civilization to promote the green transition of Chinese economy and society, so as to achieve sustainable development.

Hence, China's attention to ecological civilization takes place at a time of accelerated industrialization, and the shortcomings of industrialization are beginning to be prominent. After entering the twenty-first century, China clearly set the target of building a well-off society and included ecosystem construction as a target into the government's policy-making process. However, it did not consider the issue from the height of social civilization. In 2002, the 16th National Congress of the Chinese Communist Party included building a civilized society with good ecosystems as one of the four major targets of comprehensively building a well-off society. It decided to

> constantly enhance the country's sustainable development capability, to improve ecosystem environment, significantly improve resource utilization efficiency, to promote harmony between human and nature and to boost the entire society to progress on the civilization development path of advanced production, well-off life and good ecosystem conditions.

In 2005, Hu Jintao, the Chinese president, called for

> improving the legal and policy systems for ecological construction, making national plans for ecosystem protection and massively performing ecological civilization education in the whole society.

In 2007, the 17th National Congress of the Chinese Communist Party was held, and Hu Jintao made higher requirements for development on the basis of the target to comprehensively build a well-off society. The requirements are increasing the coordination and balance of development, expanding social democracy, enhancing cultural construction, accelerating the development of social institutions, and building ecological civilization. The Chinese president also further clarified the main objectives of ecological civilization as

> recycling economy should reach a large scale and the share of renewable energy shall dramatically increase. The emissions and discharges of main pollutants should be effectively controlled and the quality of ecological environment distinctly improved. The ideology of ecological civilisation should be firmly established among all social members.
>
> Building ecological civilization, in essence, is to a resource efficient and environmental friendly society that is based on resource and environmental carrying capacity, following natural laws and targeting at sustainable development.

On January 29, 2008, at the third collective learning session of the Central Political Bureau of the Chinese Communist Party, President Hu Jintao stated that,

> the new requirements of fully realise the targets of comprehensively building a well-off society are: China must comprehensively push forward economic, political, cultural and social construction as well as ecological civilization construction, promoting the balance and harmony of various processes and aspects of modernization, and boosting the harmony between production relationship and productivity, political and legal systems and economic foundation.

The country needs to provide clean and fresh air, fertile land, and beautiful environment for its people through ecological civilization construction, to improve development coordination and balancing, and to promote the economy's rapid and healthy development. It needs to enhance ecological democracy construction, expanding socialist democracy, and to protect the interests and rights of the people and social equity and justice. It needs to intensify ecological cultural construction, to improve all social members' ecological civilization knowledge, and to promote all-around human development. It also needs to set up ecological compensation mechanism, realize ecological justice, optimize income distribution, narrow income gaps, and boost the balanced development of different regions.

Improving Ecological Efficiency and Increase National Economic and Technological Competiveness The tide of new industries and science and technology revolution based on clean production technologies and resource recycling utilization technologies has come. Ecological efficiency, focusing on improving energy utilization efficiency and reducing environmental pollution, will be the winning point of future international industrial competition. China needs to build an industrial system of optimized structure, high ecological efficiency, resource conservation, and environmental friendliness. It needs to develop new emerging industries and increase the Chinese economy's vitality, international competitiveness, and sustainable development capability. It needs to improve the country's independent innovation capability around the core theme of ecological efficiency; to create core competiveness in the technologies and industries of clean energy, recycle economy, and environmental protection; and to take lead in international technology development. Finally, it needs to establish and improve the economic systems, governance mechanisms, and market systems for ecological economy development.

Improving Ecological Security and Protect the Country's Nontraditional Security Nontraditional security, which mainly includes energy security, resource security, and environmental security, has risen and become the main contents of national security. Enhancing ecological civilization construction is an effective measure for improving national security. China needs to increase its energy utilization efficiency, develop renewable energy, and improve national energy security; it needs to develop recycling economy, to promote comprehensive resource use, and to maintain the supply security of water, land, and mineral resources. It also needs to intensify ecosystem protection and construction, to enhance environmental governance, and to improve environmental security.

Implementing the Principle of Ecological Justice, Promoting the Harmony Development Among Different Regions, and Adjusting Income Distribution Maintaining ecological justice is to enable the people to enjoy ecological democracy and ecological welfare, to fulfill their ecological obligations, to promote harmony and balanced development among different regions, and to narrow income gaps. It is necessary to implement ecological function zoning and planning, highlight the unique ecological features of local economy, optimize industrial layout and distribution, and increase the links among the ecological economies of different regions. It is necessary to implement ecological pricing and ecological compensation, establish market mechanisms that can reflect ecological scarcity, and reduce the development level differences and income gaps because of ecological function differences among different regions and different social groups. The government needs to conduct ecological purchase and use public finance to promote ecological protection and construction.

Enhancing Ecological Civilization Construction, Advocating Rational Consumption, and Increasing the Country's Soft Power China needs to promote ecological ethics, universally spread ecological awareness, make ecological awareness national shared value, mainstream thinking and fashionable concept, and create the tides of attaching importance to ecosystems, protecting ecosystems and rational consumption. China needs to enrich the connotations of ecological civilization building, explore its traditional ecological culture, and create systematic ecological civilization systems with Chinese characteristics according to the needs of ecological civilization building. It needs to intensify the promotion and international dissemination of ecological culture, to lead international debate, to increase China's cultural soft power, and to improve its international position.

Chapter 3
Sustainable Industrialization

Industrialization is both the cause and the carrier of the industrial civilization. The industrialization process not only indicates the direction and depth of the industrial civilization evolvement but also reveals its spatial limitation and need for transformation. The industrialization stage of a country reflects its economic development level in a comprehensive way. If China is at the intermediate or intermediate-to-late industrialization stage, how much room is left for its industrial scale expansion if it follows the common industrialization path? Under the limitation of resources and environment, will the new eco-civilization development paradigm improve the industrialization process and accelerate its transformation?

3.1 The Industrialization Process

Industrialization is a process with certain directions and different stages. In the traditional agricultural society, agriculture was the dominant sector, while the commercial service sector served as the subordinate industry. After the industrial revolution, industrial production was no longer constrained by natural cycles and productive forces and enjoyed higher production efficiency, as well as faster capital accumulation and income increase. To some extent, the industry surpassed nature and gained a much higher influence in national economy. With the rapid expansion of the market, the commercial and service industries expanded in scale and gradually became the dominant sector.

From the perspective of industrial structure change, the increments of economy expansion mainly come from industries, and this leads to an increased ratio of industrial output in economy. Industry development can be based on agriculture, which provides raw materials for some industrial activities. Some industries such as the textile and food proceeding industry are based on agricultural products. These industries are labor intensive and the products are necessities. For example, the British's industrial development started with the textile industry. The economic development theory formulated by Peigang Zhang (1949) claimed an integrative

J. Pan, *China's Environmental Governing and Ecological Civilization*,
China Insights, DOI 10.1007/978-3-662-47429-7_3

industrialization that includes the agriculture sector as a part of the industries.[1] However, many manufacturing industries, such as the iron, steel, chemicals, and cement industries, and transportation and home appliance industries that appeared in the twentieth century do not need agricultural products as raw materials. Therefore, William A. Lewis (1955), an American economist, advised countries to achieve fast economic growth by industrial development, for industrial activities are not affected by natural output and can gain rapid development through increased expansion and investment.

Another American economist, Hollis B. Chenery, studied the roles and changes of various subsectors in the manufacturing industries from the long-term perspective of economic development. He divided the industrialization process into different stages based on developing economies' experiences of structural transformation from traditional agricultural economy to industrial economy. Based on criteria like the per capita income level and industrial structural change, Chenery divided the industrialization process into three stages: the preindustrial or early industrialization stage, the industrialization stage, and the postindustrial stage. Almost all of the least developed countries are in the early stages of industrialization, and all developed countries are in the postindustrial stage, while other developing countries are at different phrases in the industrialization stage.

On the relative importance of the agriculture, the industry, and the service sectors, one of the most prominent changes of industrialization is the shrinking share of agricultural output in GDP. Even in the least developed countries, their agricultural output ratio keeps shrinking as well. In some developed countries, the agricultural output ratio is as low as 1 % (Fig. 3.1). Only in the developed economies

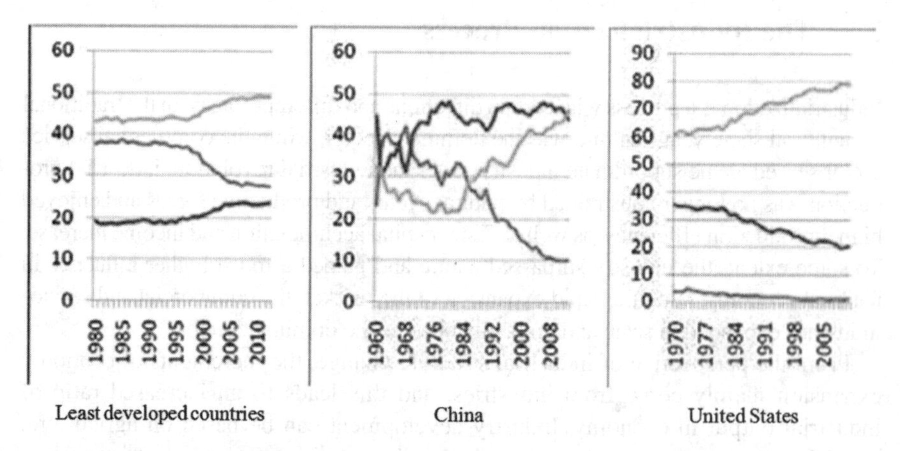

Fig. 3.1 Structure of three main economic sectors of least developed countries, China, and the United States

[1] Zhang, P. G. (1949). *Agriculture and industrialization*. Cambridge: Harvard University Press. [1949 first English edition; 1969 second English edition; 1984, first Chinese edition by Huazhong University of Technology Press; 1988, second Chinese edition].

the ratio of service sector is stable and remains at high levels (two thirds or even higher), however, for the developing economies, the ratio usually first increases then decreases. Although the industry development is the impetus of industrialization, the ratio of industrial output is usually less than 50 %, and the ratio can be less than 20 % both in the less developed agricultural countries and the most developed countries.

With regard to the service sector, in the agriculture-dominated economies, the service industries are mainly passive and subordinate to agriculture. However, in the industrialization stage, the service industries, which were derived from the secondary industries, are initiative, sustainable, and growing continuously. The interior structure of the industrial sector is the key to distinguish the stages of industrialization process. In the early stages of industrialization, the main industries are agricultural product processing and the production of primary products, such as food, tobacco, textile, mining, nonmetallic minerals such as construction materials, rubber products, wood processing, petroleum, chemistry, and coal. Most of the manufacturing industries are labor intensive.

In the intermediate stage of industrialization, the interior structure of industry shifted from fast development of the light industries to heavy industries such as iron and steel, cement, chemistry, and machinery manufacturing. In this stage, nonagricultural labors become the majority of a country's working force, heavy industries develop fast in order to satisfy the needs of mass infrastructure construction and investment, and the capital intensity of industry increases sharply. In the postindustrial stage, the service sector changes from steady growth to high-speed growth and becomes the main impetus of economic development. The service industries at this stage are not traditional service industries but emerging service industries that are capital intensive and based on advanced technologies, including finance, information, advertisement, public utilities, and consultancy.

The postindustrialization stage begins after the industrialization process is finished. In this stage, the interior structure of the manufacturing industries transits from the capital-intensive industries to technology-intensive industries. The service industries directly supporting production activities, such as industry designing, gain rapid development, leading to an increased share of the service sector in national GDP and a decreased percentage of the industry sector, which shrinks to less than 30 or even 20 %. The material products flourish and people's living standard is greatly improved.

In the postindustrialization stage, the continuous decrease of the percentage of industry is a new trend. For example, in the United States, the percentage of industry has dropped for over 15 % since the 1980s. This phenomenon where industrial and manufacturing industries experience a shrink is called deindustrialization.[2] It is the opposition of the industrialization process. This process begins at the later stages of industrialization or the postindustrialization stage and has a decreasing industrial production capacity and manufacturing activities, especially the heavy industries,

[2]Cairncross, A. (1982). What is deindustrialisation?. In F. Blackaby (Ed.), *Deindustrialisation* (pp. 5–17). London: Pergamon.

including the social and economy changes it triggers. In theory, the deindustrialization process can be a proactive adjustment, for example, when it is due to economy maturity; on the other hand, it could be some kind of passive change when it is due to loss of competitiveness. With the improving of manufacturing productivity, the relative cost of manufacture products will decrease; manufacturers will decrease their production scales through outsourcing or production activity relocation. These activities will lead to a shrinking percentage of manufacturing industries in developed countries without large adverse consequences.

Globalization and economic structure adjustment contributed significantly to the deindustrialization process in some industrialized countries. Due to the modernization of transportation and communication and information technologies, economic globalization leads to foreign direct investment, capital flow, and labor migration, and manufacturing industries move to countries or regions that are more competitive or with cheaper endowments. The space left from the outward moving of manufacturing industries is filled by expansion of modern service industries. The deindustrialization process leads to the decrease percentage of industrial sector while increase of service sector. Another consequence of the deindustrialization process is the shrinking of labor-intensive industries and jobs in industrialized countries. One reason for this is trade, especially the trade liberalization negotiation and treaties starting in the 1980s significantly speed up the moving of labor-intensive industries to developing countries where the wages and relevant standards are low. China's fast industrialization process after its reform and opening up originated from the development of export-oriented industries in the coastal provinces since the 1980s. The essence of the export-oriented economy is that material sources and markets were both abroad, and processing productions took place in coastal regions to take advantage of the cheap labor and lax regulations. Another reason of the manufacturing job loss in developed countries is technology advances such as industrial automation and digital control, which leads to higher productivity that reduce or replace many workers.

The industrialization process is also the process of material consumption reduction per unit of product. Steam engine technology in the first industrial revolution was featured by large sizes, high source and energy consumption, and high emissions. The electric motor technology in the second industrial revolution brought industrialization into a new electric era. For example, automobiles, airplanes, and ships have much smaller engines but are able to maintain the same driving power. Other mechanical equipment also dropped in the consumption of material and energy, as well as the emission amount, but was able to produce the same power. In the third industrialization revolution, computer became a common leading technology and brought industrialization into the era of information and blurred the boundary between the industrial sector and the service sector. The third revolution brought us composite materials, nanomaterials, new energy vehicles, and information networks, to name but a few. All these increased efficiency as well as lowered material consumption and emissions.

However, the decrease in material consumption for per unit of product did not reduce natural resources demand and pollutant emissions during the industrialization.

The reason is simple: the rate of production scale expansion is much higher than the rate of material using efficiency improvement. For example, one hundred years ago, one thousand people owned less than one Caron average. However, nowadays the number is tens and hundreds times higher than that, while the energy and material consumption of cars have only decreased by 50 % or 100 % at the most. In the steam engine era, people wrote letters to each other that took days or months to arrive; in the electric era, people sent telegraphs or called each other by telephone which still took time and costs money; now in the information era, people are connected by the Internet where people can communicate by emails or instant message software without any time or space limits. Therefore, industrialization is also a process where living quality is improved and living pace is accelerated.

The industrialization is also irreversible. Modern material life cannot be reversed to the preindustrial society. The consumption of nonrenewable resources, especially fossil fuel, is also one way and irreversible. It is the irreversibility that makes people worry about energy security and a sustainable future. Certain environmental pollution, especially soil contaminated by heavy metal, cannot be restored in a very long time. Industrial development, urbanization, and modern agriculture production impair ecological systems and occupy ecological space and accelerate the loss of biodiversity. Once a species is extinct, it disappears and this is hard to be undone. The mass infrastructure construction in urban areas hardens the land surface that makes it hard to be reclaimed.

Industrialization is also a process where the economic society and the environment become more vulnerable. In the traditional agriculture society, most of the natural disasters and emergent events are limited to their local areas. However, with the development of industrialization, societies and economies are much more closely connected, and information also spreads much faster. One local emergent event can be known globally immediately. On one hand, the fast dissemination of information helps a society to react immediately and effectively; on the other hand, it can add instability to a society. For example, in a global economy, an economic and financial crisis in one country can quickly spread to the rest of the world; crop failure in one place can cause the grain price instability on the world market. The "butterfly effect" tells us that any dysfunction of the transportation, communication, and power or water supply infrastructure can threaten the whole system. Any misconduct may cause huge society impacts. For example, in the city trunk road with heavy traffic, an unintentional scraping caused by a driver's negligence can lead to a traffic mess and affects hundreds or even thousands of vehicles.

According to the industrialization process, China is still in intermediate-to-late industrialization stage that needs further development. However, China's industrialization space is confronted with constraints of market demand as well as resource and environmental constraints. This means that China's industrialization process cannot follow natural evolvement; instead, it has to transit to a new development paradigm, improving the industrialization process under the guidance of ecological civilization that meets the socioeconomic needs and resource-environmental capacities.

Here three aspects of the industrialization process, scale, structure, and technology, are studied. The scale expansion of industry can increase employment and growth and bring positive impacts to social economy; thus, the central government and many local governments in China encourage the scale expansion. However, the scale of manufacturing industries is limited by the market demand and resource-environmental spaces. The production scales of heavy and chemical industries and general consumer goods have been close to their peaks since the reform and opening.

On the other hand, there is still great development space for China's industrialization even within the market and resource-environmental limits. Firstly, production capacity reaching peak doesn't mean shrinkage of production scale. On the contrary, the capital needs further accumulation to satisfy socioeconomic needs. Take the iron and steel industry as an example; in developed countries, the production remained high for a long time even after the iron and steel production peaked. At present, China will continue its investment in infrastructures and buildings for a long time in order to satisfy the needs of the urbanization and living condition improvement. Secondly, under the globalization context, China's enterprises with advanced technologies and high productivity that are competitive in global market enable themselves to keep and seize a large market share. Thus, the limit on scale space is not absolute but changeable through market competition. Thirdly, the industries can increase outputs, enhance energy efficiency, and mitigate pollution and can create new products and expand new markets if the markets are saturated. Thus, the constraints of environmental capacity are not necessarily the limits to industries and products. Fourthly, the limited space constraints of environmental capacity or resources can bring new opportunities to some emerging industries, such as the renewable energies, energy-efficiency industries, pollution abatement industries, etc. To ensure energy security, control pollution, and mitigate greenhouse gas, China needs to significantly adjust the energy structure, which includes increasing the renewable energies percentage in primary energy consumption up to 20 % by 2030 and increasing the industry scale by three to four times.

China's industrial structure in 2013 shows it is stepping into the postindustrialization stage. However, it takes long time to adjust one country's industry structure. In the 1970s when the United Kingdom and Japan had similar industrial structure, their industrial sectors' shares in GDP were higher than 40 %. The United Kingdom took advantage of its international finance and education and culture industries to adjust its economic structure. In the 1990s the percent of manufacturing industry in GDP of the United Kingdom was lower than 5–6 % than that of Japan. However, China doesn't have these advantages like the United Kingdom or Japan to develop its high-end services nor can China adopt deindustrialization like the United Kingdom. The industrial sector's percentage may decrease to 30 % by around 2050. This means that China's economic adjustment should *focus on the internal adjustment of its industrial sectors and products*.

For certain industries or products, their adjustment is closely connected to socioeconomic development and environmental situations. In 2012 the atmospheric haze pollution and its health impacts had become a nationwide concern, and the need for air freshener increased explosively. However, when the air quality attains the national standards after 10 or 20 years, the needs for these products will shrink

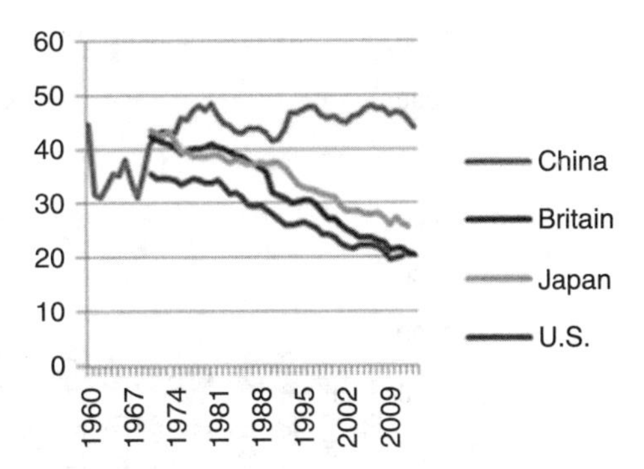

Fig. 3.2 Changes of the shares of manufacturing added value in GDP for China and several developed countries (1960–2013)

or disappear. Technology advance is the leading impetus of industrialization. The ecological civilization transition of China needs the fuel of technology innovation (Fig. 3.2).

3.2 Different Stages of Industrialization

The per capita income is one of the main criterions used to divide different industrialization stages. Many scholars have studied the industrialization levels and stages of China. According to an authoritative assessment report issued by the Institute of Industrial Economics of the Chinese Academy of Social Sciences,[3] China in general has been in the intermediate-to-late or late stage of industrialization in 2010. Cities like Beijing and Shanghai have been in postindustrialization stage, eastern coastal regions are in the late industrialization stage, and provinces in the middle and western regions are in intermediate stage, while several western provinces such as Gansu, Yunnan, and Guizhou still remain in the early stage of industrialization.

Income is a relative and dynamic process. According to per capita gross national income (GNI) levels, the World Bank divides countries in different income levels into five categories: low income, lower middle income, middle income, upper middle income, and high income. Based on the 2013 currency exchange rate, if the GNI per capita is lower than 1045 US dollars, it is a low-income country; between 1046 and 12,745 US dollars is in middle income, which covers both lower middle income and higher middle income; if it is higher than 12,746 US dollars, it is high-income country. The world average per capita GNI is 10,513 US dollars, and high-income country is 38,623 US dollars; China's per capita GNI is 6807 US dollars, in range of the higher middle-income countries. If measured in market exchange rate, China as a whole is in the intermediate stage of industrialization; however, if measured in purchasing power parity (PPP), China is already in postindustrialization stage. According to the World Bank statistics on the per capita GNI of different

[3] Chen, J. G., Huang, Q. H., Lv, T., & Li, X. H. (2012). *The report on Chinese industrialization (1995–2010)*. Beijing: Social Sciences Academic Press.

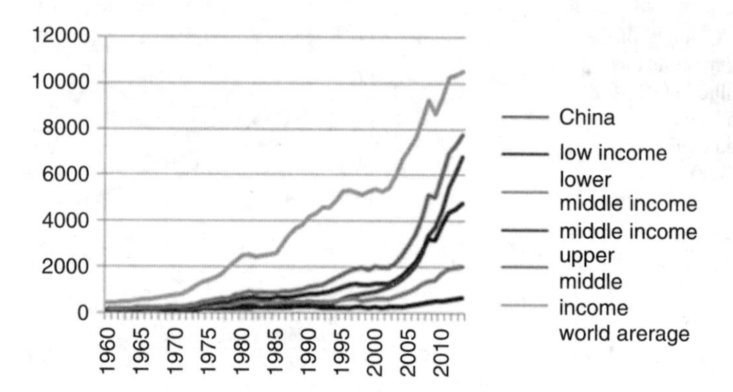

Fig. 3.3 The trend of per capita GDP (1960–2013 in market exchange rate) (Data source: World Bank. http://data.worldbank.org/)

countries (Fig. 3.3), China's per capita GNI surpassed the low-income level and became a lower middle-income country in 1984; then it surpassed upper limit for lower middle-income countries and joined the list of middle-income countries in 1995; China surpassed upper limit of the middle-income range in 2008 and is expected to surpass the higher middle-to-high-income level by 2016 at the current growth rate. The world average per capita GNI falls in the range of the upper middle income. Thus, China can begin its postindustrialization stage in 2016 if measured in PPP; however, the time will be postponed to around 2025 if measured in market exchange rate.

Table 3.1 reveals that the development of the industrial sector in China is more advanced than that in other countries with similar development levels. If measured by the criteria of agriculture and industry ratio, China had been in the early stage of industrialization in the 1970s and in the postindustrialization stage in 2013. Compared to this criterion, China's per capita GDP level is lagging behind the industrialization process, and the urbanization level is also lagging behind.

When will China enter the postindustrialization stage? The industrialization process is closely connected with economic development, urbanization, industry structure, and employment structure. The growth rate has decreased since 2013 but remains at a relative higher level. The economic growth target was set down to 7.0–7.5 % in the 12th Five-Year Plan and will slow down gradually in the future. According to projections from different research institutes, it is predicted that China's GDP will increase to 14 trillion US dollars by 2020, and its per capita GDP will be higher than 10,000 US dollars by then, which is still lower than both the world average and the high-income threshold. China's GDP will double to 28 trillion US dollars by 2030 and become the biggest economy in the world (in market exchange rate), and its per capita GDP per will be as high as 20,000 US dollars.[4]

[4]The International Monetary Fund reported in the World Economy Outlook on October 2014 that China's GDP had exceeded the United States about 200 billion US dollars measured in PPP and

Table 3.1 The timeline of different industrialization stage for China

Criterions	Industrialization stages: Before industrialization (I)	Early stage of industrialization (II)	Middle stage of industrialization (III)	Late stage of industrialization (IV)	Postindustrialization stage (V)
① GDP per capita (2005 USD, in PPP)	745–1490 (1985)	1490–2980 (1992)	2980–5960 (2002)	5960–11,170 (2009)	Above 11,170 (2016)
② The year China achieved	1985	1992	2002	2009	2016
③ The year China surpasses income levels of different groups of countries	1984 (low income)	1995 (lower middle income)	2008 (middle income)	2016 (upper middle income)	2025 (world average)
④ Economic structure the year China achieved	A>I (<1970)	A>20 % and A>I (1993, <1970)	A<20 % and A>S (1993, 2012)	A<10 % and I>S (2014, 2012)	A<10 % and I<S (2014, 2013)
⑤ Urbanization rate by population the year China achieved	Below 30 % (1993)	30–50 % (1994–2010)	50 %~60 % (2011–2020)	60 %~75 % (2020–2035)	Above 75 % (2036)

Note: ① criterions refer to Chen et al. (2007); ② the years that China achieved are measured by PPP, based on IEA (2013); ③ the data of developing countries as a whole come from the World Bank database, and the standard of per capita GDP levels is defined by the Wo rld Bank; ④ A is agriculture, I is Industry, and S denotes the services sector; the years China achieved certain structural thresholds come from World Bank database; ⑤ the data of urbanization rate by population and the years that China achieved certain thresholds are from China Statistical Yearbook

Thus, from the income perspective, China is projected to enter the postindustrialization stage in around 2025.

According to China's *National New Urbanization Planning (2014–2020)*, the urbanization rate will reach 60 % at 2020, just reaching the threshold of the late industrialization stage. According to this urbanization criterion, even with an urbanization increase rate of one percentage point per year, China will only enter postindustrialization stage after 2036. From the point of view of the practical needs of China's industrial transition, the industrialization criterion better fits to the reality of ecological civilization development. The urbanization rate criterion, which is drawn from the experience of developed countries, may not fit to China's reality. The relatively lower urbanization rate is perhaps better for the coordinated development of urban and rural areas.

China's endeavor to adjust the industrial structure is showing initial effects. The added value of the service sector had surpassed the industrial sector for the first time in 2013, signifying the locomotive of economic growth has shifted from the manufacturing industry to the service sector. In the medium and long term, the growth rate of industrial sector will slow down, and its percentage in GDP will keep shrinking. According to the domestic institutions' prediction, the industry growth rate will be as high as 8.27 % and be the main impetus of economic growth during 2010–2020. However, in the next three decades, the average annual growth rate will gradually decrease to 6.39 %, 3.80 %, and 2.46 %, respectively. The percentage of the industrial sector will decrease to 43.1 % in 2020 and 38.7 % in 2030. Based on the above analysis, the time that China as a whole enters the postindustrialization stage will be in 2015 the earliest and 2030 the latest; the confidence interval will be between 2020 and 2025.

Different Chinese regions enter into the postindustrialization stage at different times. Beijing and Shanghai are taking the lead in finishing industrialization; this is both reasonable yet for special reasons. Beijing is the national center for politics, science and innovation, and international communication, while Shanghai is the national economic and financial center. Both of them have monopoly and irreplaceable positions. Beijing and Shanghai, as the two largest cities in China, have the best resources in science, education, culture, and medical care, and they assume the role of national transportation hub. Due to their unique political and economic position, most of the headquarters of the largest state-owned enterprises and regional headquarters of international corporations are located in these two cities. In 2013, out of the 85 mainland corporations that are among the top 500 companies, 48 have their headquarters in Beijing and 8 in Shanghai, constituting two thirds of the total. Strictly speaking, Beijing and Shanghai, as the center cities serving the whole nation, are incomparable and irreplaceable in their industrial structure and urbanization level that reflect the industrialization stage.

The cross-country transfer of the manufacturing industries is the main driving force and pathway of deindustrialization for developed countries. However, China

become the number one economy entity. However, the China's government and academic society have not acknowledged this viewpoint.

is an economy with obvious regional differences. For China this transfer includes two aspects: domestic transfer between different regions and international transfer. In the eastern regions, due to the lack of land resources, increasing labor wages, and constraints of environmental capacities, many labor-intensive industries have transferred to the middle and western regions. In Beijing, besides the Shougang Group, a big steel-making company that has already moved away from Beijing, many other labor-intensive service industries have chosen to move out after the Initiative of Coordinated Development of Beijing, Tianjin, and Hebei Province in 2014.

The deindustrialization in the eastern regions will not affect the whole industrialization process nationwide. Around 2025 when China as a whole enters the postindustrialization stage, will industries start transferring abroad? According to theories like the flying geese paradigm and the gradient technology ladder and experiences from developed economies such as the United Kingdom, the European countries, the United States, and Japan, the answer is yes. With regard to China, due to its shortage of natural resources, limited fossil fuels, and environmental capacities, transferring abroad is apparently a rational choice and a potential trend.

However, this trend is limited by several factors, which determine that China's industrialization process will not follow the deindustrialization process of the developed countries. Firstly, due to the large regional disparity, there is huge space to absorb various industries. Under the background of globalization, some industries with high-energy consumption, high pollution, and high emission try to transfer abroad to avoid issues from domestic resource and environmental limitations. However, requirements of ecological civilization, regulations of host countries, and international sustainable development trend will make this kind of "Pollution Haven" activities with little marginal costs but great risks.

Secondly, China can choose the import substitution strategy to manufacture the source materials into semi-products or even final products in adjacent places. This strategy seems to be appealing. However, the main source materials, such as iron ore, dairy products, bean, and timber that China imports are from Australia, South and North America, Russia, etc. These countries have very high immigration standards and environmental regulations and have few low-end labors, which means there are no cheap labor sources and high environmental investments. All these constraints make this strategy next to impossible.

Thirdly, China has great domestic needs on raw materials for the urbanization, infrastructure construction, maintenance, and renewing and production of consumption goods. This is hard to be satisfied by the manufacturing industries of any foreign countries. China's urban population will be over one billion at around 2030, which is larger than the total population of OECD.

Fourthly, but most importantly, China needs large amount of job opportunities for its huge labor force. Even though the expanded higher education has significantly improved the quality of labor resources, the majority of employment is still in manufacturing and low and intermediate service industries. Lastly, China's demands for high-end services still mainly rely on developed countries. For example, there are over 200 thousand students studying abroad every year.[5] China import a lot more cultural products than it exports, such as films and movies. With the socioeconomic development of China, the trend will change and China will transfer from a net importer of high-end services to a net exporter. However, this process will take a much longer time than the industrialization process itself.

3.3 The Room for Scale Expansion

Based on the above analysis, the current stage and future process of China's industrialization indicates that there will still be great room for the scale expansion of its manufacturing industries. How much is it? It depends on the market demands. It is meaningless for any scale without real demands. It is also constrained by the resource and environmental limits, which also limit the scale of manufacturing industries. Thus, the process and space of industrialization will be constrained both by the real needs and resource and environmental limits.

Based on the experience from the developed countries, the expansion of production capacity of manufacturing industries is highly correlated with the income level. The scale expansion will stop or even begin to decrease once income reaches a certain level. During the industrialization process, the heavy industries and products are the symbolic industries that follow certain law of production pattern. The heavy industry products, such as the iron and steel, construction materials, nonferrous metals, and chemical industries, are also high-energy-consuming products that are mainly used for infrastructure construction and machinery manufacturing industries. These productions usually have a specific peak output, and even though there may be some fluctuations, they will not further increase with the income or industrialization process. The peak time will last for a certain period.

Due to the technology advance and energy consumption structural change, raw materials and energy consumption will keep decreasing. Developed countries that finished the industrialization process earlier have peaked their CO_2 emissions for the heavy industries during 1970–1980, when their GDP per capita overpassed 10,000 US dollars (12,000–20,000 US dollars, 2000-year price) and the urbanization rate overpassed 70 %. These criterions indicate that the completion of large-scale infrastructure construction, the scale expansion of heavy industries coming to an end, and the industries as a whole enters into the postindustrialization stage. Table 3.2 illustrates the socioeconomic conditions when the iron and steel production peaks in some developed countries. This can reflect the relationship between the peak output of the heavy industry and the industrialization process.

China's industrialization started late; therefore, the large-scale infrastructure construction and machinery production still need to consume a large amount of raw materials, such as iron and steel. In fact, the United States is the number one

[5] Center for China and Globalization. (2012). Annual report on the development of Chinese students studying abroad, No. 1.

Table 3.2 The socioeconomic conditions and the iron and steel peak outputs in chosen countries

Country	Peak time	Urbanization rate (%)	GDP per capita (USD in 2000)	Peak output lasting period	Iron and steel peak output (100 million tonnes)	Output at 2012 (100 million tonnes)
United States	1973	74	20,395	9 years	1.37	0.89
Japan	1973	74	15,531	Decreasing	1.20	1.07
United Kingdom	1970	77	12,540	10 years	0.28	0.10
Germany	1974	73	13,390	Decreasing	0.59	0.43
France	1973	73	13,787	8 years	0.27	0.25
Korea	(2012)	84	24,640	Increasing	–	0.69
China	(2013)	54	6087	Increasing	–	7.79

Sources: World annual raw steel production (1975–2009). http://xxw3441.blog.163.com/blog/static/7538362420 10112075918299/?suggestedreading&wumii World Steel Association, 2013; China National Statistical Bureau; World Bank database

producer and produced 8.33 billion tonnes of crude steel from 1871 to 2013, out of which 2.04 billion tons from 1871 to 2013, which is one quarter of the total amount. China ranks as the second in the list, producing 7.95 billion tons and 7.6 million tons (0.1 % of the total), respectively. For the United Kingdom, the numbers are 1.67 billion tonnes and 495 million tonnes (30 % of the total). From this perspective, China's basic raw materials industries still need further development. From the perspective of crude steel production, China is close to 600 kg per capita in 2013, which is very close to the peak output of the United States and Germany (Fig. 3.4). If China keeps this amount for 40 years like Japan did, its accumulated production of crude steel will come close to 40 billion tonnes. For the United States, even if it keeps up its current output until 2050, its accumulated production of crude steel will still be less than 12 billion tonnes.

According to the data from the National Bureau of Statistics, in 2013, the fixed asset investment is 44.7 trillion *yuan*; newly constructed railway 5586 km, among which the high-speed railway 1672 km; newly constructed roads 70,274 km, with 8260 km being high-speed railways; newly built cable line 2.66 million km; real estate investment 8.6 trillion *yuan*; construction area 6.656 billion m^2, among which newly construction area 2.01 billion m^2, finished 1.01 billion m^2, sales 1.31 billion m^2; vehicle production 22.12 million, among which automobiles 12.10 million; production of the electricity generation equipment 125.73 GW; 779 million tonnes of crude steel, which is nearly half of the world total amount for many years.[6] Considering the amount of the infrastructures and manufacturing industries that are formed by these huge fixed asset stock, we can hardly imagine that these levels will

[6] Data from World Steel Association, 2013.

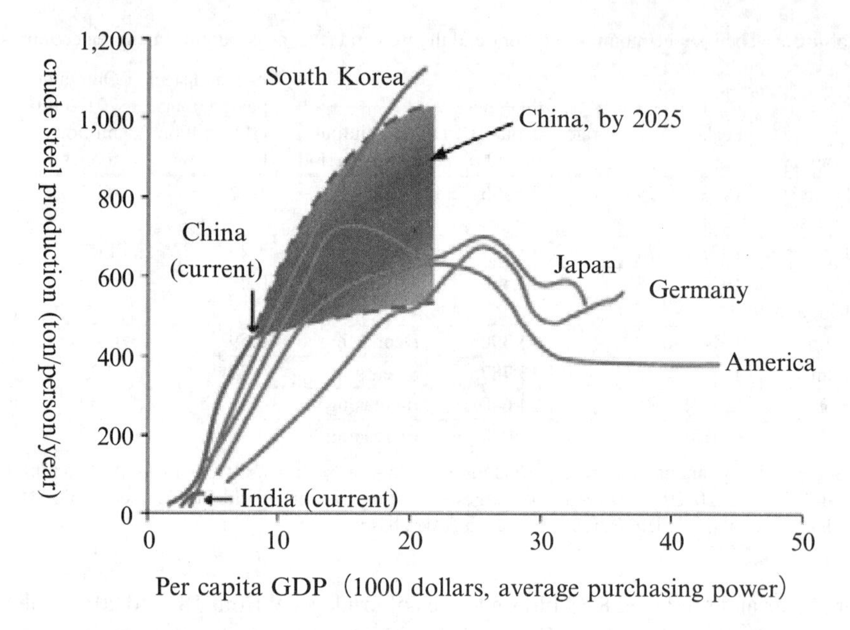

Fig. 3.4 The per capita income and per capita crude steel production in selected countries

last for another 40 years. It is not possible nor necessary.[7] In this sense, China's crude steel production has been close to or already achieved its peak output. Either measured with the per capita or total amount level, there is very limited space for further increase.

In the 2010s, the scale, competitiveness, and technology of China's manufacturing became prominent in the world economy. According to China's Statistics Yearbook, the production of iron and steel, coal, electricity, cement, fertilizer, cotton, and so on are the largest in the world.[8] On the other hand, there is limited room for the export and domestic fixed asset investment. The market demands in developed countries are becoming saturated, while the demands in other developing countries are still under development, and there is a competitive relationship between these countries and China. According to the part of the production data of durable goods from the National Statistical Bureau (2014), in 2013, the production of cell phone was 1.46 billion, LCD TV 123 million, and air-conditioning 131 million. The world population is 7.1 billion and China's population is 1.35 billion. There will be limits for the market capacity of the consumption goods even with product upgrading needs. The production capacity of some capital- and labor-intensive manufacturing industries has been close to or even surpassed their peak outputs. The industry and product

[7] BHP Billiton. (2011). Steelmaking materials briefing, presentation by Marcus Randolph, 30 September. http://www.bhpbilliton.com/home/investors/reports/documents/110930%20steelmaking%20materials%20briefing_combined.pdf

[8] China Statistical Yearbook, 2011, p. 1057.

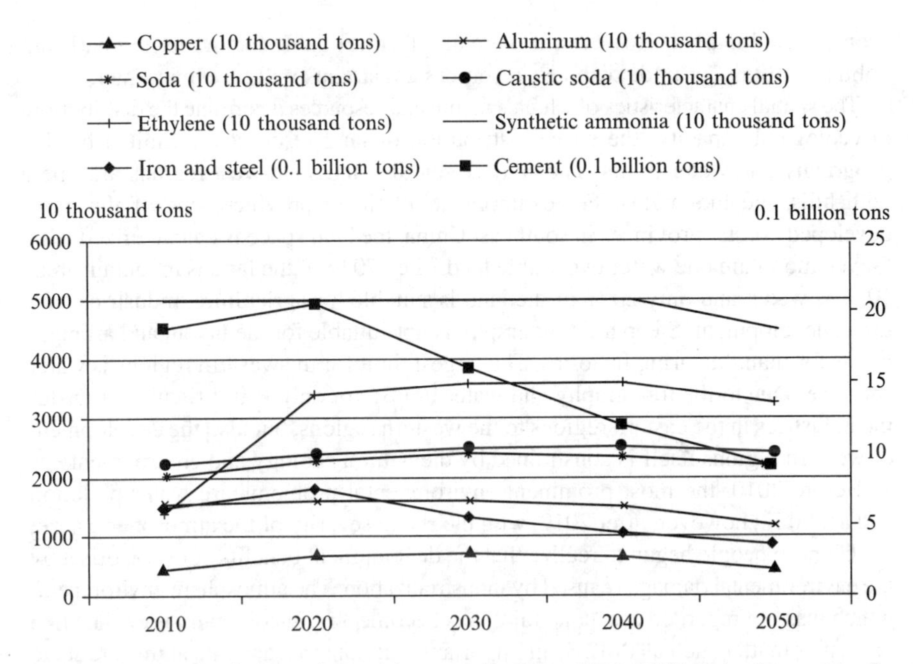

Fig. 3.5 Projected main heavy industrial outputs for China (2010–2050) Note: the numbers for 2010 are real; other years are projected (Source: China statistical yearbook, 2011; 2050 China energy and CO_2 emissions report (2009))

upgrading are a consecutive process. For example, large-displacement high-energy cars will probably be replaced by electricity cars in the future; the categories will change, and the added value will increase; however, their quantity will not increase at such high rate.

Several institutions in China had studied and predicted the industries at sectorial level. The results reveal that most high-energy and high-emission heavy industries and raw material industries will peak before 2020; after 2020 they will decrease slowly and gradually (Fig. 3.5).

China's shares in the world total metal product production, such as copper, aluminum, zinc, nickel, and lead, were only 5 % in the 1980s shortly after the reform and opening up; now they are higher than 40 %. Some international institutions predict that by 2020 they will be higher than 50 %. From the perspective of mineral cycle, these metals are not renewable; however, from the perspective of renewable economies, a large percent of the products are recyclable. There is a cost-benefit calculation on what are the percentages that can be used for recycling and what are the acceptable costs and energy consumption for these recycling. It is unequivocal that the reserves of fossil fuels have limits; however, we can produce and utilize the renewable energies. The keys are the price and quantity of the renewable energies. If the production of the renewables cannot satisfy the needs, there will be rigid constraints for these resources. The prerequisite for this is that we have an open economy and market. However, if there are uncontrollable factors in international politics, military, or trade disputes, resources constraints will be even tighter.

The spatial characteristics of China's nature and resources determine the distribution of ecological capacity; the space distribution of land resources is limited by the geography and water limits. The coastal, middle, and southwest regions are filled with hills and mountains. For example, in Zhejiang province, one of the most developed coastal provinces in southeast China, the land space is characterized with "seven mountain one water two arable land," i.e., 70 % of the land is mountain area, 10 % is water, and only 20 % of the land is suitable for agriculture, industries, and urban development. Steep hill topography is not suitable for the layout and arrangements for manufacturing factories. The largest limit for the western regions is water shortage. Due to the topography and water limits, not only is it difficult to transfer the industries in the eastern regions to the western regions, but also the development of western regions itself is constrained by the natural ecology and environment.

Before 2010, the most prominent environmental problems are water pollution and acid rain; however, after 2012, with the rising severity of the atmospheric haze, the Chinese people began to realize that the development benefits could be offset by the environmental damages caused by industrialization. The atmosphere environment, which was not regarded as a constraint, has become a rigid constraint now. The fine particular matter such as PM2.5, an important pollutant that appears at the late stage of industrialization, was not included in the pollution monitoring list until the spring of 2012.[9] However, the PM2.5 pollution has already become very serious. From the monitored results of some piloting cities, the PM2.5 concentration in most cities has exceeded the new national air quality standard (in which the threshold for the second level is 35 $\mu g/m^3$ for annual average PM2.5 concentration) and is much higher than that of the World Health Organization (WHO) (10 $\mu g/m^3$) or compared to other cities in developed countries. Furthermore, the haze pollution covered most regions in China. In urban agglomerations such as the Capital Circle (Beijing, Tianjin, and Hebei Province), the Yangtze Delta, and the Pearl River Delta region, the haze pollution durations are more than 100 days per year and some cities even more than 200 days per year. In the beginning of 2013, there was a most severe haze pollution event with record long duration, wide impact, and high intensity. In January 2013, the pollution covered 2.7 million km^2 and 600 million populations, including more than 40 key cities in 17 provinces (municipalities and autonomous regions). According to the PM2.5 data released by the Ministry of Environmental Protection in 2014, among the 74 main cities, only three cities have reached the national air quality standards (these are Zhoushan in eastern Zhejiang province, Haikou in Hannan province, and Lhasa in Tibet plateau); the PM2.5 concentrations

---

[9] The former Prime Minister Wen Jiabao chaired a State Council executive meeting and agreed to issue a revised *Ambient Air Quality Standards*, which newly included the PM2.5 and ozone pollutants with 8 h concentration limits monitoring indicators, and deployed and strengthened the comprehensive prevention and control of air pollution measures.

in 33 cities are 30 times higher than the WHO standard (PM2.5 < 10)[10] and have adverse impacts on public health, economy, and society.

The components of haze pollution match the characteristics of the current industrialization stage. The majority of PM2.5 originated from anthropology activities, which includes the primary pollutants from the coal combustion and vehicle emissions, and secondary particular matters chemically formed by the primary pollutants in the atmosphere. Due to different geological locations and socioeconomic conditions, the PM2.5 pollution sources and their contributions in different cities also vary. For example, according to the source analyzing results from the environmental protection agency of Guangzhou, the PM2.5 pollution sources of Guangzhou are 38 % from vehicle emissions, 32 % from industry pollutions, and 12 % from site dusts, and 18 % are volatile organic pollutants. In Sichuan, based on the results from Sichuan Environmental Monitoring Center, the PM2.5 pollution sources are 22–25 % from industry pollution; 16–20 % from vehicles; 17–20 % from coal combustion; 20–25 % from site dust, fumes, straw burning, paint solvents, etc.; and 10–18 % from other sources.[11] The seriousness of air pollution matches the characteristics of the late industrialization stage in the developed countries, which means that China has entered into the late industrialization stage where environmental capacity begins to limit industrial scale expansion.

3.4 The Paradigm Transition of Pollution Governance

Since the 11th Five-Year Plan period (2006–2010), based on the national pollution mitigation measures, a high percentage of the environmental capital has been invested in construction of wastewater treatment plant, installation of desulfurization and denitrification facilities in thermal power, installation of flue gas desulfurization facilities for steel plants, etc.[12] However, these environmental investments are mainly focusing on local pollution mitigation, ignoring the locked-in effects of increasing energy consumption. This brings conflicts between environmental protection and energy saving. Apparently this high-investment and high-energy-consuming

[10] According to study from ZhongNanshan, an academician in the Chinese Academy of Engineering, if PM2.5 concentration increases from 25 to 200, the inducing mortality rate will increase 11 %. Another report, *Toward an Environmentally Sustainable Future: Country Environmental Analysis of the People's Republic of China*, which was jointly finished by environmental experts from abroad and domestic and working group from the Asia Development Bank, claims that the annual economic losses due to the air pollution is as high as 1.2 % of GDP if calculated by the illness costs or 3.8 % if based on willingness to pay.

[11] The differences for different cities lie in many factors, including the limits of analyzing methods and also other uncertainties such as impacts of surrounding environment, validity of monitoring data, structure of emission sources, meteorological conditions, etc.

[12] Ministry of Environmental Protection of People's Republic of China. (2012, September 7). The announcement of national main pollutants emission reduction in 2011. http://www.mep.gov.cn/gkml/hbb/qt/201209/t20120907_235881.htm

paradigm for environmental protection under the industrialization civilization is unsustainable.

High-Investment High-Energy-Consuming Paradigm for Industrial Pollution Control China's government has increased investment in environmental protection to 2.16 trillion *yuan* during the 11th Five-Year Plan period (2006–2010), improving the inadequate investment in the 10th Five-Year Plan period (2000–2005), which is 1.58 times higher than that in the 10th Five-Year Plan period. Compared to investment needs of 1.53 trillion *yuan*, the actual investment was 632 billion *yuan* higher than the investment needs.[13]

In general, the absolute environmental investment in each urban environmental infrastructure construction kept increasing. In 2010, the environmental investment was 665 billion *yuan*, which is 1.67 % of GDP, an increase of 47 % as compared with the year before. Among the 665 billion *yuan*, investment in urban environmental infrastructure construction was 422 billion *yuan*, an increase of 68.2 %; the investment in industrial source pollution was 39.7 billion *yuan*, a decrease of 10.3 %; the investment of "three simultaneousness"[14] in construction projects was 203.3 billion *yuan*, an increase of 29.4 %.

In key areas, the investments that focus on industrial pollution reduction have also significantly increased. The investment of industrial pollution control in 2010 was 39.70 billion *yuan*, among which the investment for industrial wastewater treatment projects was 13.01 billion *yuan*, and the new design capacity was 9.07 million tons per day; the investment for industrial waste gas treatment projects was 18.88 billion *yuan*, and the new design capacity was 290.28 million m^3 per hour; the investment for industrial solid was treatment projects was 1.43 billion *yuan*, and the new design capacity was 247 thousand tonnes per day.[15]

During the 11th Five-Year Plan period, the environmental protection driven by the high investment focused on pollutant reduction and had met the absolute reduction objects for wastewater treatment and local pollutants such as sulfur dioxide (SO_2). Compared to 2005, the chemical oxygen demand (COD) and SO_2 had decreased 12.5 % and 14.3, respectively, both over fulfilled the total emission reduction targets of the 11th Five-Year Plan.[16]

[13] Lu Yuan-tang, Wu Shunze, Chen Peng, & Zhu Jianhua. (2012). Evaluation of environmental protection investment during the "11th Five Year Plan" period. *Chinese Population, Resources and Environment*, *10*, 43–47.

[14] "Three simultaneousness": according to the latest Environment Protection Act (Article 41), for the projects under construction, its pollution control facilities must be designed, constructed, and used simultaneously with the main project.

[15] Annual Statistic Report on Environment in China. (2010). Ministry of Environment Protection. http://zls.mep.gov.cn/hjtj/nb/2010tjnb/201201/t20120118_222728.htm

[16] Annual Statistic Report on Environment in China (2010). Ministry of Environment Protection. http://zls.mep.gov.cn/hjtj/nb/2010tjnb/201201/t20120118_222728.htm

However, accompanying the achievement of the pollution reduction is the raising of energy consumption, which kept increasing with the GDP growth. During the 11th Five-Year Plan period, the national energy consumption per unit of GDP had decreased 19.1 %, less than the planned 20 % target. Therefore, higher environmental protection investment is effective on pollution reduction especially for the local pollutants, but it has limited role in reducing energy consumption per unit of GDP. In the 12th Five-Year Plan period (2011–2015), China will still be in the intermediate-to-late industrialization stage; the industrialization and urbanization will keep developing at high speed, which will further exacerbate the tensions between resources, energy, and the environment.

The mass investment in environmental protection will continue during the 12th Five-Year Plan period. Premier Minister Li Keqiang announced at the opening ceremony of the High-Level Conference on China-EU Urbanization Partnership that the accumulated investment in environmental protection will overpass five trillion *yuan*, which provides huge potentials for investment and development in energy-saving and environmental protection fields.[17] According to an assessment study, the investment need in environmental protection and pollution reduction during the 12th Five-Year Plan period will be as high as 3.4 trillion *yuan*, which will effectively increase the GDP by 4.8 trillion *yuan*.[18]

In the 12th Five-Year Plan, two new binding targets are added, for ammonia and nitrogen oxides emissions. In addition another three important binding targets are also added: the energy intensity of GDP shall decrease 16 %, the CO_2 emission intensity of GDP decrease by 17 %, and the percentage of nonfossil fuels in total primary energy consumption increase to 11.4 %. These three new targets bring great challenges to the environmental protection work in the 12th Five-Year Plan period. Furthermore, there are some emerging environmental problems. During more than 100 days per year, haze covers the Yangtze River Delta, Pearl River Delta, and the capital urban agglomerations; the PM2.5 concentrations are two to four times higher than the WHO standards; photochemical smog frequently occurs.[19] The integration of regional economy, the integrity of environmental problems, and the pollution transportation between cities through atmospheric circulation, all of these bring great challenges to the current environmental management model. If the high-investment and high-energy-consumption model continues, the realization of those energy intensity and renewable ratio targets and the reduction of CO_2 emission and PM2.5 concentration will be very challenging.

---

[17] Li Keqiang. (2012, May 4). The total investment during the 12th Five-Year Plan period will surpass five trillion *yuan*. *China Daily*. http://www.chinadaily.com.cn/hqcj/gsjj/2012-05-04/content_5826884.html

[18] Lu Yuantang, Chen Peng, Wu Shunze, & Zhu Jianhua. (2012). Specifying the investment needs for environmental protection during the 12th Five-Year Plan period, ensuring the realization of the environmental protection targets [J]. *Environmental Protection*, *08*, 53–55.

[19] Chinese Academy for Environmental Planning, Ministry of Environmental Protection. (2010). Planning guide on the joint prevention and control of air pollution in key areas in the 12th Five-Year Plan Period (draft). http://www.caep.org.cn/air/DownLoad.aspx

The High Investment in Environmental Protection Combined with the Narrow Target of Pollution Reduction Will Cause Wastes in Energy Consumption and Further Pollution During the 11th Five-Year Plan period, the government adjusted the strategic deployment of environmental protection and increased the total amount of investment in environmental protection, which not only improved the investment structure but also underlined the narrow "pollutant reduction" goal. The underlying causes for the environment-energy paradox need further studying. The current fashion of environmental protection which focuses only on the pollutants reduction regardless of the cost-efficiency and energy consumption is "gray" environmental protection, which was caused by the officials' performance appraisal scheme in the environmental protection field. Under the guidance of the official promoting scheme and the narrow "pollutant reduction" goal, investments in environmental protection will be unable to attend all problems and can't change the overall situation.

Take the urban wastewater treatment as an example; under the narrow pollutant reduction goal, the urban wastewater treatment facilities are high-energy consumption industries, with low efficiency and high costs. The focus has always been on the construction and achievements of wastewater treatment plants, ignoring energy consumption during the operation process.[20] The absence of energy consumption consideration will cause unsustainability.

In the standard wastewater treatment process, the main energy-consuming processes are wastewater lifting, aeration system, and sludge treatment. The type of energy consumption includes electricity, fuel, and decontamination agents, among which the electricity consumption constitutes 60–90 % of the total energy consumption. The traditional wastewater treatment technology is highly energy intensive. It consumes enormous fossil fuels, generates large amount of residual sludge, and emits huge quantity of CO_2.

Large wastewater treatment plants are usually equipped with large and high-energy-consuming equipment. Excessive energy consumption has become a main obstacle for the development of wastewater treatment plants. The average electricity consumption for cities in China is 0.290 $kW \cdot h/m^3$; 82 % are less than 0.440 $kW \cdot h/m^3$. The average energy efficiency of Chinese wastewater treatment plants is equivalent to the levels of the developed countries at the end of the twentieth century. For example, it was 0.20 $kW \cdot h/m^3$ in the United States and 0.26 $kW \cdot h/m^3$ for Japan in 1999 and 0.32 $kW \cdot h/m^3$ for Germany in 2000.[21] Wastewater treatment requires high electricity consumption; with the increasing electricity prices, many large wastewater treatment plants often have to stop operation due to insufficient budget. These plants have already consumed large amount of environmental investment. Their low operational time has not only caused significant investment waste but also resulted

[20] Zhang Wenhao. (2012). *On the energy saving and operation technology optimizing in urban wastewater treatment plant*. Taiyuan University of Technology.

[21] Yang Lingbo, Zeng Siyu, Ju Yuping, He Miao, & Chen Jining. (2008). The statistical analysis and quantitative identification of energy consumption for municipal sewage plant in China. *Water & Wastewater Engineering*, *10*, 42–45.

in huge energy consumption. This is causing many environmental protection investments unable to create the expected environmental benefits.

Another example is the recycling and disposal of e-wastes. Under the narrow target of pollutant reduction, many investments focus on the reduction of the electronic wastes (e-wastes), ignoring the induced pollution caused by toxic pollutants from those e-wastes during the disposal process. The main methods to dispose the e-wastes are quite preliminary, such as manual dismantling, burying, and burning at low temperature (600–800 °C).

Electronic products, such as computer, TV, voice speaker, and cell phones, contain large amount of toxic substances and will produce materials with more toxins when incinerated under low temperature, such as the furans, polybrominated dioxins, and polybrominated benzo. The leachate after combustion will pollute water sources and air, cause heavy metal pollution in rivers, and increase CO_2 emissions. The landfill where the e-wastes are buried will cause leakage to the land and endanger the environment. Even the most advanced landfill approach cannot guarantee no leakage, let alone those primitive landfills with low standards.

In Guiyu town, Shantou County, the largest collection and distribution center for e-wastes in Guangdong province, there are more than one million e-wastes disposed each year. Sample tests for the bank sediments in Guiyu town showed high concentration of various toxic heavy metals, the concentration of Barium is ten times higher than the critical threshold for dangerous soil pollution, the tin is 152 times, chrome is more than 1300 times, lead is more than 200 times, and the pollutants in water are thousand times higher than the drinking water standards.[22] Such environmental investment that narrowly focuses on reducing the volume of electronic wastes has proven to cause more environmental pollution than environmental protection.

Furthermore, part of the environmental protection work is linked to the circular economy concept and invests lots of money in 100 % recycle of resources; this is a waste of extra energy and resources. The circular economy is feasible in theory, while in practice we have to consider the cost-effectiveness for the recycle of resources. One hundred percent recycling only costs more resources and capital instead of being truly energy saving and environment friendly.

The above points fully illustrate the issues on the current environmental protection practice guidelines. Thus, we need to accelerate our pace in reinventing the traditional guideline and integrating environmental and energy issues to open up a new phase in environmental protection.

Energy Saving and Environmental Protection Need to Consider the Pollutant Treatment Capacity and Energy Consumption in a Comprehensive Way Overall, there is no fundamental conflicts between energy saving and environmental protection. This has been studied and demonstrated extensively by scholars globally.

[22] Lu Yang, & Wang Yan. (2012). Status and management countermeasures of E-wastes in China. *Environmental Sanitation Engineering*, *8*, 34–36.

Taking the wastewater treatment as example again, there are many ways to reduce energy consumption while improving the treatment capacity. If China wants to keep the current treatment technologies constant, there are other ways to reduce the energy consumption: either through increasing the energy efficiency of each equipment by assessing each energy-consuming source and upgrading or through increasing the consumption of nonfossil fuels to reduce the energy consumption and emissions. Improving energy efficiency is the direct way to save energy and reduce the energy intensity target. However, the latter solution is more forward looking and strategic, not only does it help to realize the energy intensity target but also helps to reduce the CO_2 emissions, both of which can significantly contribute to the fulfillment of other targets of environmental protection.

Compared to the traditional fossil fuels, nonfossil fuels can significantly reduce the greenhouse gases (GHGs) and other air pollutants such as smog, dust, etc. There are extensive studies on how to utilize biogas and solar energy in the wastewater treatment process. First of all, the biogas produced in the anaerobic tank during the wastewater treatment process can be used for generating electricity. The biogas power generation is an integrative energy using technology that is not only environment friendly but also energy saving. It utilizes the biogas to generate power and make full use of waste heat to further produce biogas; in this way the integrated thermal efficiency can reach as high as 80 %, much higher than the traditional 30–40 % level. Thus, it is a good way for wastewater treatment with high economic benefits. Studies show that a German wastewater treatment plant not only can realize energy self-sufficiency but also can export energy[23] through improving biogas utilization from the anaerobic tank. In addition, it can improve the anaerobic technologies in order to increase the biogas production.[24] Furthermore, solar power heating and electricity-generating equipment can be introduced to replace fossil fuels during the wastewater treatment process and can make use of the untreated sewage heat by the "water source heat pump."

Besides that, there are other ways to improve the efficiency of wastewater plants, such as better site selection and size and design optimization, to decrease costs and improve energy efficiency. For example, small-scale power plants can be built near wastewater plants to avoid high-energy consumption and emissions and investment waste caused by idle equipment.

There is another way to treat the wastewater, i.e., use the constructed wetland to naturally purify wastewater. This approach can also preserve the biodiversity at the same time. This is a new technology to treat wastewater in a biological way that emerged in the last decades. It organically combines the two goals: purifying water and preserving the biodiversity. This method ameliorates the environment, creates biological landscape, and brings environmental benefits and economic benefits. The constructed wetland method has received high attention due to its unique

[23] Schwarzenbeck, N., Pfeiffer, W., & Bomball, E. (2008). Can a wastewater treatment plant be a powerplant? A case study. *Water science and technology*, *57*(10), 1555–1561.

[24] Greer, D. (2012). Energy efficiency and biogas generation at wastewater plants. *BioCycle*, *53*(7), 37–41.

advantages, and it has been used to deal with sewages, industrial wastewater, and wastewater from mining and oil drilling sites.[25]

As mentioned above, there are many ways to find solutions to solve the energy and environmental paradox, and conflict between energy saving and environmental protection is not inevitable. In the intermediate-to-late industrialization stage, in order to achieve energy and environmental goals, it is needed to "green" the environmental protection practices when there are large-scale investment in pollution control and environmental protection. There are many ways to reduce the energy consumption, such as increasing energy efficiency, increasing nonfossil fuels, considering the ecological effects, and making full use of solid and gas wastes. This will not only need extensive research but also need to change the narrow target of pollution reduction and make better use of the investment resources.

Change Strategy and Include Both the Energy Saving and Emission Reduction into Targets in Environmental Protection Practices The traditional high-investment and high-energy-consuming strategy needs to be shifted to a comprehensive strategy. From an overall perspective, energy saving is more in line with the trends of sustainable development than the environmental protection. Firstly, in order to fulfill targets in energy efficiency, CO_2 emissions, and nonfossil fuels, China needs to combine energy saving and environmental protection and apply a multidimensional appraisal scheme for the environmental protection practices. Furthermore, we need to improve the monitoring and establish a sound system of the environmental monitoring indicators. In 2013 the Ministry of Environmental Protection (MEP) had initiated a monitoring scheme for six indicators (including PM2.5) in 113 cities and released the monitoring data in December 2013.[26] This is a positive signal. And further improvement is needed in the monitoring capacity for the extreme climate events and emerging environmental pollutants in the future. Thirdly, we should increase the nonfossil fuels, mitigate emissions, and promote "green" environmental protection. Nonfossil fuel in China has fast development rate and large investment scale but needs higher investment efficiency. Thus, we need to adjust the current investment strategy and use public fund to leverage private capital to develop nonfossil fuels.

---

[25] Huang Jinlou, Chen Qin, & Xu Lianhuang. (2013). The problems and solutions of the constructed wetland in practice. *Environmental Science*, *01*, 401–408.

[26] Zhou Shengxian's address on the national conference on environmental protection at January 24, 2013. http://ml.china.com.cn/html/guanzhu/20130125/134515.html

Chapter 4
Harmonious Urbanization

China's urbanization is of enormous scale, lasts a long time, and has far-reaching influences. In the early 2010s, China's urban population has already been more or less the same as that of the European Union (EU). It is projected that China's urban population will be nearly or equal to the total urban population of the Organization for Economic Cooperation and Development (OECD) countries and far exceed the total urban population of all the high-income OECD member countries. The transition from rural economy to urban economy is a civilization transformation. The development paradigm of industrial civilization won the competition with and replaced the development paradigm of agricultural civilization, which was of low productivity and passively adapted to the nature, leading to expansion of urban wealth and increase in the population-carrying capacity of cities that can meet the needs of social and economic development. In 1978, the total urban population in the high-income OECD countries was 610 million; in China it was 180 million. 35 years later, the former increased to 860 million, with a net increase of 250 million; in China it grew to 720 million and the net increase was 540 million. If China's urbanization rate reaches the OECD average level, its urban population will see another increase of 400 million. The urbanization process in the developed countries had generated huge impacts on the world economy and environment. Impacts of the urbanization in China will be magnified under the background of economic globalization. Plus with the huge population and short time frame of the process, China's urbanization will not only have significant impacts on the Chinese economy, society, and environment but also generate significant and far-reaching influences on global sustainable development. If China does not adjust and transform its urbanization paradigm, not only will China's urbanization process encounter various obstacles, it will also cause high social, resource, and environmental risks.

J. Pan, *China's Environmental Governing and Ecological Civilization*,
China Insights, DOI 10.1007/978-3-662-47429-7_4

4.1 The Urbanization Process

At the global level, the urbanization under the industrial civilization paradigm does not face much technology and investment constraints, but it encounters enormous resource and environmental constraints. The ultimate goal of urbanization is human development. This is especially evident in China. The urbanization in China is for the better life of the people, the social welfare of both urban and rural citizens is at stake on it, it involves economy upgrading and transformation and development paradigm change, and it concerns energy security, climate security, and environmental security. Thus, China needs to go beyond the narrow technical issues in its urbanization policy making; it should define and regulate the country's urbanization process from the broad macroeconomic perspective of society, economy, resources, and environment.

4.1.1 The Duality of China's Urbanization

China's urbanization level was only 11 % in 1949, and when the reform and opening up started in 1978, it was 17 %, going up only 6 percentage points during the three decades. During the early 1960s, China's urbanization rate reported some negative growth, and during the "Cultural Revolution," tens of millions of junior middle school and senior middle school graduates left the cities to work and live in the rural areas due to lack of jobs in the cities. Because of the enormous differences in urban and rural income level, social welfare, and development opportunities, the Chinese government resorted to the administration of *hukou*, a resident registration system, as an instrument to curb the rapid growth of urban population, creating the urban and rural dual system in government administration. However, after the reform and opening up, even though the *hukou* system does not change, the control on the free movement and migration of population has been gradually relaxed. In the beginning, rural labors were allowed to "leave their field but remain in the rural areas"; this has gradually changed into hundreds of millions of farmers migrating to cities for job. During the process, China's urbanization has been progressing steadily and at high speed. Since 2010, China's urbanization rate has been above 50 % and it keeps increasing at 1 percentage point each year. The *National Plan on New Urbanization (2014–2020)* released in 2014 estimates that China's urbanization rate will increase to 60 % by 2020. Various studies predict that the urbanization will continue and reach 70 % around 2030, which is equivalent to the urbanization level of the moderately developed countries; that means a net increase of 300 million people in China's urban population. Due to constraints of the dual *hukou* system, among the existing 720 million urban population, there are 260 million rural migrant workers who live and work in cities, but do not enjoy the same access to local urban public services and social welfare as their urban counterparts.

If the urbanization during the first 30 years since 1978 is a kind of "passive" urbanization propelled by the export-oriented industrialization, the urbanization in the 2010s is a kind of "active" urbanization aimed at boosting domestic demand through eliminating discriminations against migrant workers and giving them the same rights of accessing to public service and social welfare as other urban residents. The strong domestic demand drives industrialization and leads to active urbanization. By 2030, the urban population will see another increase of 300 million on the current basis. A total of 560 million people, as new urban residents, will need to have access to urban infrastructure facilities, public service, and social welfare; this is bigger than the total population of the 28 EU countries. This means each year, China needs to provide another 25 million new city residents with urban infrastructures and social services. The lasting and massive urbanization and integration of new urban residents will become the huge and sustained engine for China's economic growth and upgrade.

Statistics from the United Nations Development Programme (UNDP) and the World Bank indicate that the average urbanization level of the high Human Development Index (HDI) countries is over 80 %; in the upper-middle HDI countries, it is 74 %; among the lower-middle HDI countries, it is 34 %. In 2013 China's HDI was ranked the 91st in the world, in the middle level. Cities such as Beijing and Shanghai have already been included in the list of high HDI cities. There are strong correlations between a country's urbanization level and its HDI level. Improving the quality and level of urbanization is a huge practical challenge and a strategic task for future social development.

To a large extent, urbanization depends on fossil fuels for providing energy for city construction and operation. In 2010, the total population in the OECD countries was 1.23 billion; among them one billion live in urban areas, and the total energy consumption of OECD countries was 5.41 billion *toe* (tons of oil equivalent). China's total energy consumption was less than 0.5 billion *toe* in 1978, and it increased to one billion *toe* in 2000 and further to 2.47 billion *toe* in 2010. These numbers reflect the growth rate and development trajectory of energy consumption during the urbanization process. It is predicted that the Chinese population will be 1.4 billion by 2030, among which about 1 billion will be urban citizens. If it is assumed that China's per capita energy consumption in 2030 will be the level of OECD cities in 2010, China will need 6 billion *toe* to meet the energy demand of its 1.4 billion people in 2030. The latest Fifth Assessment Reports (AR5) of the Intergovernmental Panel on Climate Change (IPCC) further confirmed that anthropologic CO_2 emission is the main factor for global warming and the world needs to significantly reduce its CO_2 emission. Since 2010, China's CO_2 emission has been over 1/4 of the world total, and its per capita emissions nearly reaches the EU level. The main cause for the nationwide haze pollution is fossil fuel combustion. To safeguard national energy security, climate security, and environmental security, China's urbanization has to follow a low-carbon path at present and in the future.

4.1.2 The Connotations of Harmonious Urbanization

China's environment and resources endowments are low and cannot support the current extensive urbanization pattern with high energy consumption and high pollutant emissions. The challenges of land, energy, and water resource shortage and environmental deterioration form the multiple rigid constraints for the scale, level, spatial distribution, and path choice of urbanization. It is impossible to build cities in the deserts, which do not have water; cities, regardless of their sizes, cannot function normally in the absence of modern energy services. It is impossible to fundamentally improve cities' sustainable development capacity simply through energy-saving and emission reductions. At the strategy and macro-social economic development level, it is necessary to realize the overall transformation of civilization development paradigm.

Firstly, It Is Necessary to Optimize the Spatial Arrangement and Scale Structure of Different Cities Since the 1990s, in the course of urbanization, large numbers of Chinese population have flowed to such megacities as Beijing, Shanghai, and Guangzhou, and regional center cities, leading to "Matthew effect" of urbanization: large cities expand faster than the smaller ones, leading to "polarizing" development of megacities. The increasingly serious "urban disease" in these cities indicates that the excessive expansion of megacities causes lack of economic scale in city development and neglect to the boundary constraints to city development. The population scale, land use, water use, energy consumption, and housing price in these megacities and super large cities have already surpassed these cities' environmental and social carrying capacities. These megacities and super large cities absorb most of the financial capitals, occupy the resources for small city development, and intensify the inequality between urban and rural, between eastern and central and western regions, and among large, medium, small cities. These megacities and super large cities, which find it difficult to serve the needs of their residents, as national or regional centers, have to provide such services as higher education, cultural, and medical services to people outside their own jurisdictions, causing such problems as traffic congestion, environmental deterioration, severe water scarcity, and overconsumption of energy.

Data from the annual *Environment Protection Communique* indicated that in 2013, China put into effect the new *Ambient Air Quality Standard*[1] and started to monitor and assess the annual average concentration levels of SO_2, NO_2, PM10, and PM2.5 in 74 cities in the key regions of the Beijing–Tianjin–Hebei, the Yangtze Delta, and the Pearl River Delta regions, municipalities, province capital cities, and cities specially designated in national development plans. Only in 3 of these 74 cities that are relatively small and located far away from densely populated areas, Haikou in Hainan Province, Zhoushan in the East China Sea, and Lhasa in Tibet, met the new standard for air quality. The other 71 cities, or 95.9 % of total of 74, failed to meet the new standards. In the Beijing–Tianjin–Hebei region, among the days

[1] Ambient Air Quality Standard (GB3095-2012).

when air pollution exceeded the new standard, during the two thirds of the days when air pollution exceeded the new standard, the PM2.5 was the most important pollutant, followed by the PM10 and O_3, which accounted for 25.2 % and 7.6 % respectively. Due to the oversize and inappropriate urban planning, the traffic congestion in the megacities is extremely serious. The average commuting time, the time people spend traveling to their work by whatever means, in megacities such as Beijing, Shanghai, Tianjin, Shenyang, Xi'an, and Chengdu, is more than 1 h. In Beijing it is as long as 1.32 h.[2] Based on a survey on vehicle speed in megacities all over the world conducted by the United Bank of Switzerland (UBS), the road traffic conditions are the best in London, with an average vehicle driving speed of 29 km per hour (km/h) in downtown areas; this is followed by New York and Singapore, both cities having a driving speed of 24.9 km/h. The average vehicle driving speeds in Chinese cities are much lower than in other countries. Particularly in Beijing, it is only 12.1 km/h, one of the lowest in the world; other Chinese cities are a little better, the average driving speed is, respectively, 16.3 km/h in Shanghai, 17.2 km/h in Guangzhou, 18.0 km/h in Chengdu, 20.0 km/h in Hong Kong, and 20.4 km/h in Wuhan. Only urbanization with proper spatial arrangement and scale control of cities can facilitate the localization of social services in the cities and reduce and eliminate the "urban diseases" due to "polarization" of resource allocation to different cities.

Secondly, Follow the Law of Nature Scientific development requires the urbanization process to follow the law of nature, be in line with the ecological balance principle, and match the local environmental capacity, climate capacity, and the resource carrying capacity. It is also criteria of judging whether the urbanization process is low carbon or not. For example, in some arid or semiarid areas in the upper and middle reaches of the Yellow River, water shortage is very severe; however, the local governments have built many man-made landscapes with mountains and waters, wetland parks, and man-made green spaces by extracting a large amount of water from the Yellow River and underground water. It seems that these cities are "green" and "ecological," but in fact they are high carbon. To build and maintain these man-made landscapes on local barren urban land, they have to use high-pressure pump to supply water and use enormous energy. Such actions are against the law of nature and not low carbon. A lot of big Chinese cities compete to build the highest skyscraper on Earth, for example, Changsha City even plans to build a "sky city" building as high as 838 m. At first glance, such actions seem to be a highly intensive and efficient use of land; however, the energy consumption and environmental impacts during the building, operation, and maintenance of such extremely high skyscrapers are much higher than those of buildings of moderate height. To achieve harmonious urbanization, cities need to thoroughly analyze their local strength and weakness, keep their development within the "redline" thresholds of local ecological capacity and resource carrying capacities, and make use of local specialties and existing resources for the development of relevant industries.

[2] Institute of Social Science Survey, Peking University, and Zhilian job searching (2012).

Thirdly, Focus on Human-Oriented Urbanization Profit-driven industrialization and urbanization work against sustainable employment, cause environmental pollution, deny the equal rights and interests of rural migrants as local urban residents, enlarge the income gaps between urban and rural residents, expand city scale by urban sprawl, but do not embody the guiding principles of human-orientation and benefit the public, and impede the healthy development of cities. During the 11th Five-Year Plan period (2006–2010), the industrial energy consumption increased from 1.595 billion tons of standard coal equivalent (tCe) to 2.4 billion tCe, making up 73 % of the total energy consumption in China. Six energy-intensive industries, including iron and steel, nonferrous metal, building materials, chemicals, and power industry, which found their share in the total industrial energy consumption had increased from 71.3 % to 77 % during the 5 years. However, in developed and mature economies, industrial energy consumption is only around 30 % of the national total energy consumption, while the transportation and building sectors, which are associated with living quality, respectively, account for approximately 30 % and 40 % of the total national energy consumption. China's current energy consumption structure in the current fast industrialization and urbanization stage indicates not only the long and difficult urbanization process ahead but also the urgency to transit to a harmonious urbanization that focuses on improving living quality of the general public. Urbanization is a historical task for achieving the modernization goal of China; it also provides enormous potential for expanding domestic demand. The principles of harmony between human and nature and harmony between individuals and the society have to be effectively implemented during the entire urbanization process. To do this, the following measures are necessary: vigorously developing modern service industries, pushing ahead industrial restructuring, changing the main driving force for economic growth from investment to consumption, improving the quality of urbanization, and avoiding overemphasizing on industrialization rate and GDP growth.

Fourthly, the Urbanization Should Support the Environmental and Living Quality Improvement in Small Towns and Villages and the Autonomous Participation of Rural Residents In the past three decades, China's "city construction" movement due to the urbanization that focuses on developing big industrial enterprises and building big cities has swiped out many small towns and villages that had beautiful natural environment and were low carbon and ecology friendly. In big cities, population density is high, and economic activities are intensive and concentrate in small pieces of land; it is difficult to realize and maintain the harmony between human and nature. However, cities can still create and maintain a livable environment through various ways, such as protecting historic sites, increasing green spaces, and creating gardens on roofs. Even if China's urbanization level increases to 70 %, there will still be around 400 million people living in the countryside, and these people need to access the same social services and enjoy the same living quality as urban residents. Improving the environment and living quality in the countryside and supporting the autonomous participation of farmers can better address the needs of rural residents. Small towns and villages have their own advantages, such as suitable for the application of distributed energy, the development of low-cost and high-return green industries, and the operation of small and diversified organic farms. This kind of urbanization can meet local needs with local supplies; it is a model for really low-carbon, environment-friendly, energy-saving, circular economy, and harmonious urbanization. Such harmonious urbanization supports the development of green and well-off life in small towns and villages and can make the countryside more attractive.

4.1.3 The Transformation of Urbanization

To choose the path for the transformation to harmonious urbanization, it is necessary to consider the different technical options, such as offering migrant workers the same treatment and welfare as local urban residents, energy saving, renewable energy, and economic stimulus. However, all these technical choices must be based on the path choice at the strategic level, so as to fundamentally realize the transformation toward the development paradigm of ecological civilization.

Energetically Boost the Upgrading and Optimization of Industrial Structure Given that China's urbanization is driven by industrialization, industries, to a large extent, are the backbone and main components of cities; therefore, the harmonious urbanization must include the low-carbon transition of industrial structure. In China, low-carbon development of cities covers both the upgrading of existing industries and the optimization of new industrial productivity. First of all, cities have to restrict the excessive expansion of energy-intensive and high-emission industries and vigorously support the development of low-carbon emerging industries, especially such strategic emerging industries as energy efficiency and environmental protection industries, new generation of information technologies, biology, high-end equipment manufacturing, new energy resources, new materials, as well as automobiles using new energies, so as to fundamentally reduce the emission of carbon dioxides. Secondly, cities need to upgrade their traditional carbon-intensive industries and lower the carbon intensity of their existing industries. Currently, the traditional industries of "high energy and resource consumption, high emissions, high pollution, and low return" still dominate the Chinese economy; therefore, renovating and upgrading the traditional industries are critical tasks for realizing low-carbon development in the cities. Thirdly, Chinese cities should speed up the development of a modern service sector and boost the carbon intensity decrease through economic structure optimization. Statistics show that the average carbon intensity of the industrial sector is around five times that of the service sector. Therefore, enlarging the share of service sector in urban economy and reducing the share of the industrial sector can reduce cities' dependence on the carbon-intensive industrial sector during their development and is an effective approach for industrial structure upgrading and boosting low-carbon economy development in the course of urbanization.

Promote the Transition of Consumption Pattern and Lifestyle Harmonious urbanization needs the ecological civilization transition of the urban citizens as well, including changes in such aspects as their consumption patterns, lifestyle, social norms, and local development rights and interests. The consumption and daily needs of urban residents include nine aspects: clothing, food, housing, using, transportation, health care, education, social relations, and entertainment. Among them, the material consumption for meeting the first five needs depends more on the quantity of resources and environment; while the other four needs are more dependent on the quality of resources and environment. The transformation of urbanization must be accompanied with the transition of lifestyle and consumption patterns, in order to make the material consumption rational, healthy, of high quality, and in conformity with the carrying capacities of local resources and environment and to keep a harmonious relationship between the spiritual needs and the nature and local ecosystems.

Strengthen Intensive, Intelligent, and Green Development An intensive resource-using pattern and compact city layout are basic requirements of low-carbon cities. Intensive land use can not only reduce the occupation and waste of resources but also realize the mix use of land for various purposes and contribute to the rejuvenation of urban vitality, the promotion of public transport policies, and the successful implementation of some ecological measures in urban communities. All of these effects can reduce the disturbance and occupation of nature and resources, keep some natural and ecological elements in the cities and integrate them in city development, and safeguard the harmonious development of the cities. The intelligent technologies and devices can provide support for the harmonious and low-carbon development of cities. The Internet and the Internet of Things can integrate urban resources. The intelligence engineering facilities can quickly process and analyze data and realize real-time management and control of people, devices, and infrastructures, especially in such public sectors as transport, energy, commerce security, and health care. The green buildings can effectively contribute to resource conservation, environmental protection, and pollution reduction; meanwhile they can also provide healthy, comfortable, and high-efficient spaces for work and living. Green transportation can transport system construction and maintenance costs and energy consumption through building coordinated transport system with low pollution and facilitate city environmental diversification. Thus, the green architecture design at microlevel and green transportation design at meso-level are two key starting points for the design of harmonious and low-carbon cities at the macro-level. Therefore, intensive, intelligent, and green design makes it possible to scientifically and properly solve such issues as natural resources, housing conditions, transportation, working environment, and leisure spaces in the course of urbanization, so as to save resources, protect environment, and reduce pollution as much as possible and facilitate the progressing of harmonious urbanization.

Implement Scientific Planning and Include Harmonious Elements in the Overall Planning and Designing of Cities China's basic profile of natural resource endowments and climate capacity determines that Chinese cities cannot select the same planning and development path of the US cities, which sprawl in space are of low population intensity. Actually the US urban development is a typical pattern of high energy consumption and high-carbon emissions, which was formed due to the abundant environmental capacity in the United States, and it has locked the cities into a high-carbon development path. In order to avoid such high-carbon locked-in effect, China needs rational urbanization development planning that integrates the idea and principles of ecological civilization and takes into account the constraints of climate capacity and carbon emission budget and builds low-carbon and resilient cities with strong natural restoration capability. Low-carbon and resilient cities are such a kind of city that can, in the course of city governance, planning, and design, combine the needs for greenhouse gases (GHGs) emission reduction and climate risk management, following the approaches for adaptive management to achieve the goals of ecological integrity and sustainable cities. To build low-carbon and resilient city, we need to transform traditional urban management pattern and ideas to synergy management of various targets, policies, and means.

Enhance Public Participation in the Making of Rules and Regulations The core of harmonious urbanization is the harmony between human and nature. In the transformation toward harmonious urbanization, the wide, active, and authoritative participation of citizens are indispensable, and it is also necessary for the urban residents to adjust and change their ideas and behaviors. In the absence of wide public participation, it will be difficult for the good ideas to be widely disseminated and accepted and even more hard to mobilize wide participation in actions. Such participation should be active, rather than passive, so as to generate personal motives for change in ideas and actions and make the transformation a conscious action of the public. The wide and activity participation in the transformation policy making can make the policies more authoritative, reduce conflicts and problems in the transformation process, and increase the harmony of transformation actions.

4.2 Sustainable and Livable Cities

Large urban population and cluster development are two main characteristics of urbanization in China, which are the inevitable results of social, economy, and environmental development under the constraints of natural resources distribution and ecological carrying capacities. The large-scale and cluster development of cities, to a large extent, can help improve the sustainable development and livability of Chinese cities; meanwhile, they also pose significant challenges to the sustainable development of the society, economy, environment, and the quality of human settlement. The governance of megacities urgently needs coordinated innovation in order to improve the overall quality and levels of cities' sustainability and livability.

There are rational basis and intrinsic driving forces for the formation of megacities in China during its urbanization process. The *Hu Line* (*Hu Huanyong Line*) that divides China into two parts with far different population densities was formed due to China's topography and precipitation distribution. It indicates that the northwest half of China's land territory is of low environmental capacity and low population-carrying capacity; the southeast part of the country, despite their mountains and hills, has relatively higher environmental capacity and population-carrying capacities. Due to China's huge population and highly intensive social and economic activities, the high concentration of population becomes an inevitable choice for China's social, economy, and the environmental development.

From the perspective of economics, megacities have the advantages of imperfect competition, increasing return to scale, and comparative costs. Information asymmetry and monopoly power on resources further increase the advantages of these megacities, leading to continuous expansion of these megacities and further enhance their increasing return to scale. Livable cities are simply cities where people live and work in peace and contentment and good living environment with sound ecosystems. The physical infrastructures such as water supply system, transportation system, and buildings offer material basis for high livability in cities; the facilities for such social services as education, culture, health care, commerce, and sports provide basic social services for cities. The economic activities in cities create numerous stable job opportunities. The natural and man-made landscapes in cities often have aesthetic, cultural, and historical background and values that can make people feel happy and relaxed. People yearn for working and living in cities because cities have livability advantages.

The scale economy of cities leads to sustainable advantages. The first advantage is the intensive utilization of natural resources. As many intensive production and consumption activities concentrate on limited urban space, the efficiency for land, water, and energy use is high and the economic returns are bigger. For example, space heating usually needs to consume lots of energy, but in large cities, space heating can be based on district heating and the cogeneration of heat and electricity, as well as residual heat from industrial process. In this way, the heating cost and energy consumption for each square meter of floor area are much lower in large cities than in small towns and rural areas.

Secondly, it can reduce transaction costs. Cities, especially large cities, are the centers for information gathering, production, and dissemination; therefore, information access is fast and cheap and the information is accurate and reliable, because the high concentration of human activities in super large cities, social activities, culture transmission, freight transportation, and daily commute are of short distance, low costs, and fast speed. Thirdly, it can increase the scale economy. The operation of large hospitals, schools, and shopping facilities needs a big local population. Community libraries in small cities are usually small and their service contents and level are limited. That's why large sports facilities and culture and arts organizations are usually located in large cities, especially in megacities. Even solid waste incinerations must be supported by a certain population size for producing sufficient municipal waste. Fourthly, large cities have innovation advantages. Cities, due to their enormous information, advanced science and technologies, and more sophisticated markets, are the sources and centers for science and technology innovation. Therefore, it can be concluded that cities have a positive role in promoting sustainable development.

City is supposed to make life better; however, more and more "city diseases" are occurring and deteriorating. This makes many urban dwellers feel confused and reconsider the benefits of living in cities and begin yearning for the idyllic life in the countryside. Apparently cities, especially megacities, are not natural phenomena, nor can they be created and maintained in traditional agricultural society. In Chinese history, the first city with over one million populations was *Chang'an*, the capital of the *Tang* Dynasty. Its prosperity lasted only a short time, because of, apart from civil wars and political factors, its huge population exceeding the local resource and environment carrying capacities and the insufficient material support for its livability and sustainability. However, the industrialization process driven by the industrial revolution has constantly been overcoming the technical bottlenecks for city size expansion. But when the city size expansion exceeds technological capacity or overpasses the rigid natural capacity constraints, "urban diseases" constantly occur, which, if not properly treated, will only become worse and worse.

The first challenge for livable city is the vulnerability to risks. Larger cities often are bigger in land area and of higher population density, huge sunk capital stock, and strong connections among the various elements in the city system. As a result, bigger cities tend to have higher social, economic, and environmental vulnerabilities. For example, the rainstorms and floods in cities are becoming more frequent, and the damages are becoming higher. During the rainstorms and floods, large areas in cities are under water, streets become rivers, people are trapped at home, and supply of drinking water, food, and electricity is cut off. Such cities can hardly be called livable. Even a rainstorm whose precipitation level is once a decade or once every two decades makes people terrified. Skyscrapers are especially vulnerable to power outage, whether it is because of such natural factors as typhoon or earthquake, or technical problems. Without electricity, skyscrapers are hardly livable.

The second challenge is environmental pollution. Nature endowed human race with blue sky, green land, and clean water. However, human production and living activities have damaged the ecological environment in cities, causing frequent haze and gray sky, futile land, polluted water, waste hills, and exhaustion of water resources.

The third challenge is the excessive shrinkage of human living space. Human beings, as part of the nature, all need certain living space. Living in the forest of concrete buildings, even if one can afford an apartment in a tower building, one owns no land and still lacks security. Living in a small apartment can be very depressing, especially if the apartment is northward apartment and can never have direct sunlight indoor. Overcrowded subway, congested traffic, shortage of beds in hospital, and difficulties to get a place for one small child in a kindergarten, all these make people very stressed, exhausted, and frustrated.

The fourth challenge is the widening income gap and social inequality. On the one hand, in cities, there are very rich people who have access to the best resources, enjoy beautiful environment, lead a luxurious life, and have various privileges, while on the other hand, a much bigger disadvantaged group lives at the fringe of cities and lacks the necessities for substance and a decent life.

The challenges to the livability of cities are also challenges to sustainable development. If a city does not have sufficient risk resistance capabilities and the disasters are irreversible, obviously the city is unsustainable. If the environmental pollution exceeds the self-purification capacity of the environment, the natural productivity will degrade, the biodiversity will decline, and people will lose the environment they rely on. The increasing income gap seems to be a social income distribution problem; actually it has negative impacts on sustainable development. Rich people waste resources and consume lots of nonrenewable resources; poor people lack the resources to protect the environment and improve efficiency. Both situations pose challenges to sustainable development. The challenges mentioned above are gradual or short-term ones; there are some long-term strategic challenges that can undermine the existence and operation of cities.

The first challenge is the energy security. City is a fruit of industrialization and the energy foundation of industrialization is fossil fuels. However, the reserves of fossil fuels are limited and nonrenewable. China's annual energy consumption was less than 600 million tonnes of standard coal equivalent (*tCe*) before the reform and opening up; however, it was over 3.6 billion *tCe* in 2012, having surpassed that of the United States and becoming the biggest energy producer and consumer in the world. However, the ratio of China's proven reserves of fossil fuels to its annual consumption is low. China's proven oil reserve is only 11 times its current annual consumption, meaning that the domestic oil reserve can only sustain another 11 years of national consumption at current level. Even though China has more coal reserve, on per capita basis, it is still lower than the world average.

And most of China's coal reserve is located in the arid regions and deserts in northwestern China; the mining, utilization, and transportation of coal are seriously constrained by local water shortage. If the fossil fuel reserves are depleted, all the industrial production activities, transportation, heating, and air-conditioning in cities will lose their energy supply and be severely affected. In the foreseeable future, it is very difficult for the quantity and quality of the renewable energy production to meet the needs of normal social and economic activities in cities. Due to the relatively small land areas of cities, the area of and energy from solar radiation are far from enough to meet the energy needs for the basic economic functioning of cities. Due to the intermittency, variability, availability, and costs of wind power, hydropower, and geothermal, their energy generation is small and unstable, making them unable to satisfy the huge energy needs of big cities. The second challenge is climate change. The IPCC Fifth Assessment Report concluded that, the global average temperature had increased 0.85 °C, the global sea level rise had reached 19 cm. If the sea level rise continues, as most of China's megacities and super large cities are located in the coastal regions, they will face daunting adaptation challenges. The temperature increase and extreme weather events due to climate change will intensify and constantly deteriorate the human livability and sustainability of Chinese cities.

The competitiveness and scale of economy advantages of the city have changed into challenges under the constraints of technology bottlenecks. Livability and sustainability are the basic criteria for harmonious city. However, the cities face not only the short-term problems of "urban diseases" but also the long-term threats to their existence. When facing these immediate difficulties and long-term threats, the only way out is social governance innovation.

First, it is necessary to promote scientific urban planning and construction. With the increase of city sizes, the urban infrastructure systems, including road, transport systems, water supply and drainage systems, and electricity and telecommunication system, also become more complicated, and their servicing capacity does not increase linearly with their scale. For example, the typhoon *Fitow*, landed on China on October 5, 2013, made Shanghai "a city in the sea," overflooded the West Lake in Hangzhou City, and caused 11.9 billion *yuan* of losses, two casualties, and one person missing in Ningbo City. The huge impacts of typhoon *Fitow* on the large cities indicate that in cities land surfaces are covered by cement and the natural water storage spaces are occupied and the floods have nowhere to flow and remain in the cities, causing lots of damages. In small cities, function zone can improve city livability; however, if megacities designate different functions for different zones, it will only lead to the creation of "sleeping cities" and traffic congestion.

Second, the resource distribution should be scattered. Due to the centralized governance hierarchy in China, cities closer to the power centers enjoy greater monopoly on resources, and the driving force for city size expansion is stronger. All the best social resources are distributed around the center of power. Polarization not only exists in resource distribution to different cities in a region but also among different areas in a city. If the provincial cities and megacities continue using their administration power to monopoly economic resources, the "urban diseases" in these cities will inevitably keep exacerbating.

The third innovation is to set high standards and make enforcement more stringent. Cities are power centers. In the absence of high standards and strict standard enforcement, the privileged groups will try all means to maintain and strengthen their embedded interests, and it will be impossible to entirely eliminate the "urban diseases." What are really missing in the social and environmental governance of cities are not laws, regulations, and standards but the authority of the law and the respect to the law.

The fourth innovation is transparency and public participation. The Internet is a good example of public participation in a flat governance scheme under the framework of laws and regulations. By contrast, close operation and hierarchy in administration are inefficient; more fundamentally, the general public refuse to agree and accept government arrangements and are reluctant to cooperate and take actions according to government requirements.

The fifth innovation is to rely on the science and technology innovation to improve governance and technology efficiency. Saving energy, water, electricity, land, renewable energy utilization, circular economy, and low-carbon development all rely on technology and governance innovation. However, it is also necessary to identify and prevent spurious technology innovations, for example, Changsha City's plan to build an 838 m high building, which will be the highest in the world. Such a building is truly land-saving but to transport water, people, and goods up the 838-m-tall building, lots of energy need to be consumed to operate and maintain the building. It is impossible for such a building to be energy efficient. Moreover, people who live and work in such a highly dense and closed space will feel isolated from nature and even feel frightened of being far away from nature. What Chinese cities need are genuine livable and sustainable innovations.

The sixth innovation is the unification of equity and efficiency. Equity and efficiency are two principles and criteria for the allocation and use of economic resources, the environment, and nature resources. Rich people not only have money but also occupy and use the best economic and environmental resources. This is a rational resource allocation from the perspective of market efficiency. However, it is inequitable, and therefore such efficiency is unsustainable. Therefore, the social governance of cities must include effective income redistribution and economic stimulation instruments, such as levying progressive tax on energy and water consumption, giving subsidies to consumptions that are harmonious with natural ecology status, and charging market price for basic consumption while levying high progressive for luxury and wasteful consumption.

4.3 Granting Rural Migrant Workers Full Access to Urban Social Services

China's industrialization has absorbed huge numbers of rural migrant workers moving into cities for job opportunities and a better life; however, the social services that cities offer to the rural migrants are far from sufficient to satisfy their needs. Therefore, there is a big gap between the statistical urbanization rate and the actual urbanization rate based on the share of population who has full access to urban social services. This has become a big barrier for harmonious urbanization. The urbanization rate calculated based on the number of Chinese people who lived in cities for more than six months in a year had reached 52.6 % in 2012; however, the real percentage of Chinese people who have urban *hukou*, recognized as urban residents in the official personal identification registry, and can have full access to urban social services, was only 35.3 %; the gap between the two rates was 17.3 percentage points, equivalent to 234 million people.[3] It is predicted that the nominal urbanization rate will reach 60 % by 2020; the real urbanization rate will be 45 %. The *National Plan on New Urbanization*, on the basis of further analysis, stated, the obstacle for the transition from the nominal to real urbanization, i.e., giving rural migrants the same rights as registered urban residents, is the cost, which needs to be shared by different stakeholders, including the rural migrants themselves. On the other hand, the rural migrants move into cities because they can gain some benefits. Thus, it is necessary to conduct a cost-benefit analysis on the issue of giving migrant workers full access to urban social services.

[3] Data are from the *National Plan on New Urbanization (2014–2020)* that was issued on March 2014.

Giving rural migrants full access to urban social services is a big social project with great historical significance for China's economic transformation and upgrading. In principle, the decision-making on this process, like other projects, must be based on cost and benefit analysis. If the benefit for this project is far lower than its cost, it will be a losing project. Even if it is started with administrative interventions, it will be doomed to fail and cannot last long. On the other hand, if the benefit is much higher than the cost, but the government refuses to take action or even blocks the process, it will lead to economic losses or even social instability and impede long-term economic and social development.

The existing studies on providing rural migrants full access to urban social services mainly focus on its costs and ignore its social benefits. Such studies can mislead the public, causing the wrong government decision of maintaining or even strengthening the *hukou* system that causes significant social inequality and blocking social reform. If the benefits are taken into account, it can be found that the benefits of changing rural migrants into real urban residents in government administration are huge and lasting, much higher than the costs. Correct analysis and calculation of the costs and benefits of giving rural migrants equal rights as urban residents in government administration is necessary for the society to select the right path toward a healthy, high-quality, and sustainable harmonious urbanization and realize the "Beautiful Chinese Dream."

4.3.1 Who Needs to Be Given the Equal Rights and Interests as Urban Residents?

According to Chinese *hukou* system, if a person moves into a city not because of a top-down arrangement but personal decisions, whether he or she finds a job in the city or not, the person will be denied the equal rights to urban social services as local citizens. However, in the urbanization rate calculation in official statistics, the criterion of judging is whether or not the person lives in a city for over 6 months during a year. In this way, among the people living in cities, apart from the people who are recognized as urban residents in *hukou* system, there are various people without urban *hukou*, such as the local native farmers in "villages inside cities" whose land has been recently changed into urban land use due to city expansion, and people who are not recognized as local urban residents, such as the rural migrant workers (also known as migrating population or transferred agriculture population[4]). The people without local urban *hukou* are denied the same rights as the *hukou* citizens, such as the rights to vote and to be elected, the equal employment rights,

[4]Before the 18th CPC (Central Party Conference), they are called migrant farmer workers, indicating that their identity is farmer, but their occupation is worker; or they are migrating population because they don't have *hukou* and, they cannot settle down but have to keep migrating. In the 18th CPC report, former President Hu Jintao used the term transfer agricultural population to describe this giant and special group.

children's rights to 9-year compulsory education in local public schools, health care, unemployment benefits, and pension. In some cities, especially in megacities, there are many formidable discriminations to their residents without local urban *hukou* in housing purchase, children enrollment in public kindergartens and schools, and social security.

Strictly speaking, neither all the non-*hukou* urban residents want to be officially recognized as local urban residents and have full access to local urban public services nor is it mandatory from the perspective of focusing on people's livelihood and realizing social justice. Firstly, the residents of "downtown villages" in city jurisdiction have lost their land and their livelihood security in connection with the land. They need to be officially recognized and accepted as urban residents and get full access to the social security in order to reduce the social conflicts and maintain social stability.

Secondly, for the rural migrant workers, it is necessary to distinguish the different types and groups. If they are immigrants or migrant industrial workers who have stable job and fixed living place and pay various local taxes and charges, they should be entitled to urban social services and other equal political, economic, and social rights. However, if they are seasonal workers who just come to cities for some extra income and do not plan to settle down in cities, they are not interested in being officially recognized as local urban residents in the cities they work. For example, in Xinjiang, there are many seasonal workers who come to Xinjiang for a few months in a year for cotton picking work, and some Chinese people go working abroad under labor dispatch service contracts. These people do not expect nor hope for being officially recognized as local residents and authorized the same rights as local people. Moreover, there are some people on short-term work in a city; they may stay in a city for over half a year, but they do not belong to the city. What they are interested in are mainly economic rights and interests, not some social and political rights and interests.

Thirdly, literally, the "migrant population" refers to the social groups that are not rooted in a place or do not intend to or plan to settle down at where they currently live. For example, foreign labors, permanent personnel of international agencies, and the permanent personnel of government agencies or organizations from other Chinese cities and regions, for example, permanent personnel of enterprises stationed in a city, are all migrant population. This group can rely on their employers to arrange and make sure that they enjoy the same or even better social security services than local urban residents. However, they do not have rights to vote in their host cities. They can actually satisfy their needs for urban social services through purchasing them. Compared with them, the rural migrant workers have no one else to rely on and are in disadvantage position in enjoying economic, social, or political rights; they need to buy but cannot afford basic social services.

Fourthly, the transferred agriculture population literally is a group of people who have transferred from the agriculture sector to nonagriculture sectors, and their living place has changed accordingly in the process. They are no longer engaged in farming activities and have moved from rural areas to cities. This transfer is irreversible and they have no plan to move back to rural areas. They have lost their original social, economic, and political rights in their hometowns, but are denied the various rights and interests in the host cities due to existing Chinese laws and regulations. What they need are the basic benefits and rights local urban residents are entitled to, including urban social services, equal employment opportunities, and other political rights.

The above analysis indicated that the social groups who need to be recognized as urban residents and gain full access to urban public social services and social security benefits are the native farmers who lost their land and the group of transferred agricultural population living in city jurisdictions. Among the migrant farmer workers or migrant population, the above two types need to be officially recognized as urban residents in the cities they live and authorized the same rights and interests; other migrant farmer workers or migrant population do not need this change.

4.3.2 *Objective Understanding About Benefits*

Urbanization is both the carrier and driving force of China's economic development and social progress. Its benefits are significant and sustainable in the process of social and economic transformation and ecological civilization construction.

Firstly, eliminating discrimination against migrant workers and officially recognizing them as urban residents is the driving engine for growth and upgrading of Chinese economy. Before the reform and opening up, the capital accumulation for economic growth was to some extent relying on the "dual-price system" between the industrial and agricultural products. The prices of agricultural product were artificially fixed lower than their actual values; while the prices of industrial products were intentionally fixed higher than their actual value; agricultural products and industrial products were traded at unequal values. China relied on agriculture-subsidizing industry and rural area-subsidizing cities to push ahead industrialization and urbanization. After the reform and opening up, the main driving forces for rapid industrialization and urbanization have been export and investment. Now export and investment are approaching saturation, and the potential for further growth is very limited; the main driving force for economic growth should be eliminating discrimination against the migrant population in cities. According to the *National Plan on New Urbanization (2014–2020)* launched in 2014, the population that needs to be officially recognized as urban residents and given full access to urban social services and social security will reach 136 million during the coming six years; on average each year, 23 million people will be recognized as urban residents and given full rights of urban residents. In addition, the large number of people living in "villages inside cities" will be covered by urban social services and given full access to urban social services, as well as induced demand; because of the narrowing of the gaps between rural and urban residents, these will become the huge and lasting engine for China's economic growth and upgrade. In the coming two decades, improvement of industrialization and urbanization in China has to rely on giving all the people living in cities the same treatment of urban residents.

Second, giving all people living in cities the same treatment as urban residents can bring about huge social benefit. Socialism with Chinese characteristics should eliminate personal identity discrimination, rather than consolidate or reinforce such discrimination. The basic interests of a group who makes great contribution to and their society are also indispensable part of the interests of the whole society. Identity discrimination against the over two hundred million rural migrant workers causes psychological twist, physical harms, increased survive pressures, and lack of voice for the migrant workers. Such discrimination makes it hard for them to contribute to the society. Because of these discriminations, the migrant workers have to leave their children behind in the rural areas and taken care by grandparents or other relatives. These children are the future of China as well. It is every human's unalienable right to having a place to live, having access to health care, receiving pension during their old age, and making their voice heard. From this perspective, the social benefits from treating all the people equally, no matter they are rural or urban residents, are even more than the economic benefits. The best choice for a society that is behind the "veil of ignorance" is to maximize the benefits of the disadvantaged groups.[5]

Third, equal treatment of all people living in cities as equal urban residents can bring about enormous environmental benefits. The size and changing speed of the resource and environmental pressure caused by China's urbanization and industrialization are unprecedented. The nationwide haze pollution from the winter of 2012 to the spring of 2013 made people to rethink: if the "Beautiful Chinese Dream" is blue sky, clean water, and green land, why have we polluted and damaged our environment and then dreaming and calling about going back to the beautiful natural environment we once had? Some people point out that the per capita energy and resource consumption of an urban resident is higher than a rural resident. This is true under the current urban–rural dual economy structure and confirmed by statistics. However, with the development of economy, and progress of society and technology, a compact city is much more resource efficient and environment friendly than scatted villages. The situations in developed countries can prove this.

The above simple analysis shows that there are great economic, social, and environmental benefits for equal treatment of all people living in cities as urban residents, which is the driving force and safeguards and conditions for the upgrade of Chinese economy.

4.3.3 Scientifically Analyzing the Cost

Under the background of urbanization acceleration and increasing problems because of social discrimination against migrant population in cities, some leading research institutions and think tanks have conducted some surveys and calculated the cost of equal treatment to migrant workers as urban residents. The *China Development*

[5] Rawls. 1(971). *A theory of justice*. Cambridge: Harvard University Press.

Report issued in 2010 by the China Development Foundation estimated that giving migrant workers equal rights as urban residents would cost 100 thousand *yuan* per person. While in the report of "Estimation on the Cost of Giving Migrant Farmer Workers Full Urban Resident Rights and Benefits" published in early 2013, the State Council Development Research Center of China estimated that the cost would be 80 thousand *yuan* per person. Although these studies have their grounds, underling theory, and methods, their estimates are still open to discussions, need further improvement, and even contain some flaws and errors.

From the perspective of methodology, all these costs assessments ignored the social cost. Due to the identity discrimination against rural migrant workers, in the city where they work and live, their children are denied the rights to be enrolled in the public schools and the equal rights to higher education and employment opportunity. Not only the discrimination hurts the dignity of discriminated population and makes them lack of basic security, it also causes enormous waste of human capital that is hard to estimate. In economic sense, these assessments also need to be modified. First of all, from the whole society perspective, no matter where a person lives, some investment in basic infrastructure and social security is always necessary. Although such investments are currently lower in the rural areas, it does not mean such investments are dispensable; in the past and currently, the investment is low, but it does not necessarily mean this can remain low in the future. Actually in recent years, the Chinese government is using fiscal funding to enlarge the coverage of subsistence allowance and society security to all people, although there are some differences in different regions. So the costs for equal treatment of migrant workers as equal urban residents should only include the incremental costs, not the total costs. Secondly, there is transfer payment in giving the migrant workers full rights as urban residents. The land for urban expansion came from farmers; thus, the costs of giving migrant farmer workers equal rights as urban residents should not include extra land obtainment costs. The transfer payment of urban land acquisition should be deducted from the total cost.

At macroeconomic level, the cost of giving migrant farmer workers equal rights as urban residents is a kind of investment. Urban infrastructure construction is a kind of public investment, and like other investment, it has the multiplication effects and can create job opportunities and fiscal income. Similarly, the expansion of urban basic social services, including education, health care, pension, and so on, can cost the government some money, but more importantly, it can create new jobs and improve the life quality of the general public. From the above analysis, the costs need to be scientifically analyzed to prevent partial estimation and exaggeration.

4.3.4 Breakup the Vested Interests

It is an undisputable fact that the benefit of eliminating discrimination against migrant workers in cities is higher than its cost. However, if resistance of the vested interest groups who are protected by the current institutional setup and regulations cannot be overcome, the process of giving migrant population full rights as urban residents will be hard to move forward. These vested interests exist in many ways, for example, the division of people into those "inside of the system" and entitled to urban social benefits and social security and those "outside the system" and denied access to urban social benefits and social security is derived from the urban–rural dual *hukou* policy and the dual system of privileges to state-owned enterprises and discrimination against private enterprises. The incumbent interest groups who are in powerful positions and have vested interests in the existing dual systems want to maintain and strengthen these dual systems, while the disadvantaged groups whose interests are hurt under the existing dual systems have little power and are helpless and unable to change the existing dual systems. Many cities do not mind wasting a lot of resources on constructions, destruction, and reconstructions, but are reluctant to invest in building kindergartens, preliminary schools, and community hospitals. City authorities can buy farmers' land at very low cost or even no cost and gain huge profit from land price increase, but farmers cannot get any of the profits. The migrant workers pay income taxes in their host cities, and the companies they work for pay urban maintenance and construction tax[6] and education surcharge[7]; their children should enjoy the rights of education in the host cities. The salary of rural migrant workers should include the living costs of the migrant workers themselves, which include not only the their own basic needs such as clothing, food, and transportation but also income security for the period they cannot work due to old age, sickness, injury, and pregnancy; this guarantees for the survival of disadvantaged groups. The salary should also include the costs of the reproduction of the labor force, i.e., the cost of raising their children. Some city decision-makers focus on minimizing "cost" to their cities, but ignore human "rights" of migrant workers or the social "benefits" of inclusive development. Such practices are accepting and supporting the current vested interests structure. However, such partial urbanization is accepting the contribution of migrant workers but denying them social benefits and social security; obviously it is inharmonious and unsustainable and cannot realize the Beautiful Chinese Dream.

Chinese cities can learn from the experiences and practices of developed countries in their urbanization process, which guaranteed equal opportunity and basic security. Among the Chinese who stay in the United States after finishing their study, except for a small share working in universities and national research institutions, most choose to work in private enterprises or set up and run their own business. There is hardly any discrimination against them because they are foreigners. Take housing as an example. The urbanization rate for Japan was only 27.8 % in 1945; it increased

[6] On February 8, 1985, the State Council issued the *Urban maintenance and construction tax regulations, People's Republic of China*, which began effective since 1985. While it was kept in and adjusted during the taxes reform in 1994. Due to this regulation, the tax rate is 7 % for taxpayers in the urban areas, 5 % in the county and towns, and 1 % in other areas.

[7] For all the units and individual taxpayers who pay the production tax (later as consumption tax), value-added tax, and business tax, they should pay the education fees. The additional education tax is 3 % of the production tax (consumption tax), value-added tax, and business tax.

to 72.1 % 25 years later, in 1970; the speed of urbanization was high. In order to solve the housing problem for migrant workers, Japan built lots of public housing using pubic fund. In early 1960s, Japan implemented a scheme of "local diversion" to encourage people and capital flow from large cities back to their hometowns and to promote the local employment and local urbanization. From its independence in 1965 to the early 1980s, Singapore finished its industrialization and urbanization in around two decades; the Singaporean Housing and Development Board provided apartments to 80 % of its population. In the 1990s, the proportion of people living in apartments provided by the Housing and Development Board even increased to 90 % (Chin 2004). Given the small land area and very high population density of Singapore, the policies of government monopoly housing provision and privatization not only guarantee that every family has a place to live but also stimulate its national development (Wong and Xavier 2004).

Both Legislation and Law Enforcement Are Needed to Overcome the Resistance of Vested Interest Groups If cities try to maximize the vested interests of the incumbent interest groups and ignore the interests of the rural migrant workers, then it is impossible to maximize the national and social interests. In order to overcome the resistance of vested interest groups, it needs to give the rural migrant workers both the rights and the venues to speak out their opinions. However, the National People's Congress, as the highest policy-making authority in China, did not have a few rural migrant worker representatives until the beginning of the twenty-first century; at the annual conference of the National People's Congress in 2013, there were only 31 rural migrant worker representatives; obviously they cannot represent the economic and social rights and interests of the 260 million rural migrant workers.[8]

Before the reform and opening up, there was only a single dual urban–rural interest system; after the reform and opening up, there are multiple dual systems, including the dual system of urban *hukou* and rural *hukou*, the dual system of urban *hukou* and rural *hukou* inside the jurisdiction of a city or town, and the dual system of local urban *hukou* and urban *hukou* from other cities. However, in essence they are all dual system of urban and rural residents. In order to break up the existing vested interest structure, first of all, it is necessary to clarify and confirm the social costs and social benefits of granting migrant workers the same rights and social benefits as urban residents in law, making sure that benefits of reform and development should be shared among all city residents who have contributed to industrialization and urbanization, regardless whether they have lived in cities for decades, or new city residents, or even rural migrant workers living in the city but do not have local *hukou*. During the policy making, it is necessary to recognize both the costs and social benefits of eliminating discrimination to migrant workers in cities. Second, it

[8]The rural migrant worker delegates increased from three in 2008 to 31 in 2013. Source: Yao Xueqing, Focusing on the 31 rural migrant worker delegates who represent 2600 million rural migrant workers. People's Daily, March 12, 2013.

is also important to safeguard market-based and decentralized allocation of the social and economic resources by market through legislation.

Megacities and provincial capital cities attract large numbers of migrant workers, suffer many "urban diseases," and find it difficult to support their large population. The main reason is the centralization of political power, leading to the monopoly position of megacities and provincial capital cities in the allocation of economic and social resources. In China, the high-quality education, health care, culture, and sports facilities and social services should decentralize and scatter from the megacities and provincial capitals to other large cities and medium and small cities, to increase the development vitality of secondary large cities, and small and medium-sized cities, and increase employment. Such measures can also reduce the population and transportation pressure in megacities and provincial capitals and promote justice and equality in social public resource allocation. Third and the most important is the enforcement of law, instead of selective enforcement of law. China's legal system is well developed; there are labor law, compulsory education law, and social security which all clearly indicate the equal rights of all Chinese citizens. However, some cities and decision-makers selectively enforce the law or evade their obligations to enforce some laws, leading to the ineffective enforcement of the law. For example, if the economic benefits of reserved state-owned land can be used to build affordable housing for the migrant workers, it can effectively solve the fund-raising and financing problem. The Constitution entitles all citizens the rights to vote and to be elected, but migrant workers need to exert their rights in their host cities.

4.4 Urban Planning

New urbanization is the carrier of the ecological civilization, while the ecological civilization development is an effective criterion for judging whether the urbanization is "new" or not. The spreading and exacerbation of "urban diseases" in the course of urbanization indicates the insufficiency of ecological civilization construction. President Xi Jinping emphasized the important leading role of urban planning in the city development when he talked about how to treat the "urban diseases." Scientific planning can bring about huge benefit, while wrong planning can cause huge losses; frequent direction change in planning should be avoided. Thus, during the new urbanization for transforming toward ecological civilization, scientific planning is a key factor determining whether a city can achieve harmonious development.

The Functions Should Match the Urban Spatial Pattern The spatial patterns of urbanization are sharpened by the nature factors and industrial investment, which need to be considered in scientific planning. The industrial needs for expansion and technology progress enable human to build cities and change the nature. Large investment can lead to the building of a new city in a short time. Industry expansion can enclose tens and hundreds square kilometers of land and transform it into industrial parks. Under industrial civilization, all business activities are profit-oriented and seek the fast accumulation of a large amount of capital and the maximization of profits while ignoring the basic elements of ecological civilization, such as the harmony between human and nature, respect to nature, and human-oriented development. As a result, the focuses of urbanization planning are industry, instead of city, and profit and tax maximization, instead of people's living standard and quality, causing imbalances in urban planning.

The overall urban spatial pattern in China includes the regional pattern of eastern, middle, and western cities; the scale pattern of large, medium, and small cities; and the different function-zoning pattern in cities. With the Chinese economy being increasingly integrated into the world economy after the reform and opening up, massive industrial investment is a strong driver for the urbanization process. The current urbanization patterns in China include dense city belts in the east, singular and stretch of cities in the middle, and scattered cities in the west.

The eastern coastal region has the location advantage for developing labor-intensive and export-oriented industries. The industrial expansion absorbs many migrant workers gathering there; as a result, city belts are formed. Due to the excessive concentration of migrant workers and insufficient urban infrastructures and social services, hundreds of millions of migrant workers cannot be accepted as local residents and enjoy the same social benefits and social security as local people in the host cities. In the middle region, many big industrial enterprises, especially those important industries that were moved to the inland in the 1950s, move to the eastern region in the tide of market economy. Many talented young people and funding also flow to the east. The provinces in the middle region increase the scale of their capitals in order to promote the economic ranking of their capital cities. As a result, the corrections between cities are becoming closer and stretches of cities appear. In Western China, because of the large-scale development and outward transportation of energy and mineral resources, cities develop and expand, and they are separate points on the map.

From the perspective of China's urban scale structure, the big cities have strong tendency for further expansion; the middle cities' development spaces are depressed, while small cities lack the driving force for development. The share of big cities in total urban population has increased from 24 % in 1978 to 43 % in 2013, while the share of urban residents living in the small cities dropped from 65 % to 45 %. In urban function zoning, urban planning emphasizes big industrial parks, ignoring the integration of different functions, leading to function mismatching of different zones. For example, an industrial park often covers an area of tens or even hundreds of square kilometers, and many industrial parks are located far away from urban public service facilities. Private companies construct residential buildings in stretches and mass scale in the absence of consultation and coordination with the providers of urban public services; the private housing companies often ignore or even oppose the integration of industry and residential building and the construction of relevant public service facilities.

The imbalance of city patterns and structure is the main cause of the regular large-scale migratory flow of Chinese population. In the holidays, especially during the Spring Festival, many people leave the eastern region to join their families in the middle and western regions, from big cities to the middle and small cities and rural areas. The morning and evening tides of commuters between the huge residential communities and industrial parks and downtown areas cause serious pressures on intercity and inner city transportation systems. The industry parks and population are highly concentrated in the big cities, which occupy large area of fertile land and cause shrinkage of green land and high housing prices. To grow their economy, medium-sized and small cities attract energy-intensive and high emission industries with cheap resources, especially cheap land, causing shortage of resources, water pollution, air pollution, and widespread haze. The imbalance of urban spatial pattern is one of the main roots for the occurrence and deterioration of urban diseases.

The traditional urban planning was investment driven and profit oriented; it has changed China into the "world factory." The new urbanization needs to improve and transform the industrial civilization through applying the ecological civilization ideas, scientific planning, blending cities into the natural environment, and focusing on local residents' livelihood in city development.

We Should Define the Spatial Boundaries for City Development The urban planning in the industrial civilization focus on such basic elements as technology, capital, and interests, and it is unnecessary to set the spatial boundary for city development. As long as there is profit, cities are ready to expand their spatial boundaries. However, the new urbanization under ecological civilization accepts the spatial boundary of urban planning and city development, emphasizing the integrity of city and nature and requiring the clear definition of space boundary for city development.

In fact, the city distribution patterns in the three Chinese regions indicate that the development of city is subject to the constraints of nature and environment and that there exists rigid boundary constraint. The overall city distribution patterns are in line with the distribution patterns of resource and environment. In the eastern region, the ecological and natural productivities are much higher; while in the western region, the ecology and natural environment are much more vulnerable. However, in the future urbanization process, Chinese cities need not only to provide urban public services and social security benefits to the existing over 200 million migrant workers who are already working and living in cities but also receive and accommodate another 300 million new migrant workers. Given that the existing city space distribution patterns are already approaching or exceeding the carrying capacities of resources and environment, how do we make sure that despite the rapid and large-scale urbanization, the city spatial distribution patterns are in line with the resources and environmental capacities?

Firstly, China must strictly safeguard the "redline" minimum area threshold for arable land to guarantee food security. Urbanization changes the surface of land and leads to irreversible change in land use. The productivity of each hectare of land in East China is several times to several hundred times higher than that in Western China. If the urbanization in East China is a kind of urban sprawl and uses too much land for city expansion, the low productivity of land in Western China makes it difficult to make up the land productivity use due to arable land reduction in East

China through expanding arable land in Western China. China cannot count on the world market to feed its 1.3 billion people. Therefore, in the city belts in East China, some land must be reserved for grain production and green space in urban areas. Simply safeguarding the redline minimum area threshold for arable land is insufficient; China must protect the natural environment and maintain the grain output of its breadbasket regions. Companies and cities appropriate fertile arable land for construction and urban use because they can get economic benefits through it; protecting fertile arable land must also include some economic incentives. It seems that it is a good investment to occupy fertile arable land for urban sprawl; however, from the perspective of ecological civilization, such practice is actually unsustainable and causes some losses. Take Beijing as an example, the basic supplies for its 23 million population mainly rely on industrial technologies: its energy supplies rely on the natural gas supply from Xinjiang through the West–East Natural Gas Transmission Project and the coal supply from Inner Mongolia and Shaanxi Province; its water supply depends on the water from central China through the South–North Water Diversion Project and the water from regulating the supply to its neighboring areas; all its vegetable consumption depends on railway and road transportation from other parts of China. Fossil fuels are unrenewable, the water supplies that come from water sources over 1000 km away are subject to natural fluctuations and some uncertainties; the storage and transportation of vegetables increase energy consumption and costs and increase the risk of food security. In this sense, the urban sprawl in Beijing not only occupies arable land for growing vegetables and grain crops but also causes negative impacts on preserving arable land in other regions.

Secondly, China must set the ecological redline threshold for city development based on the calculation of environmental capacities. Western China covers a vast land area but lacks water resources, which is a rigid constraint for city development. China's *West Development Program* does not mean massive urbanization, or building high pollution industry parks, or developing garden cities in Western China. It is unsustainable and against nature to pump up underground water, intercept water flows from rivers, and invest in seepage control facilities in order to build man-made wetland landscape and golf course. Furthermore, Western China and Middle China are the natural defense and water sources for East China; the ecological degradation and pollution in the Western China and Middle China will damage or ruin the environment and ecological carrying capacities in East China. This means that China's strategic city distribution pattern is not necessarily and cannot be even distribution of cities, especially in the regions with vulnerable environment in Western China; it is impossible to exceed local climate capacity in order to develop large city clusters and build large cities and local economic growth powerhouse. The city development in Middle China and Western China must respect nature and strictly keep human activities under the rigid constraints of ecological redline threshold. The city boundary is actually the constraints ecological redline threshold. The high population density, high economic activity intensity, and high pollution in cities have to rely on the support of environmental capacity in their surrounding areas. The more a city depends on the outside world, the higher its vulnerability is. "Small is beautiful," its underlying logic is to accommodate the nature and be in harmony with the nature.

Thirdly, let the cities be blend into the nature and sail with the wind. Making people enjoy the beautiful natural environment in their daily life and love their hometown is not only about the symbols of life quality, but more importantly, it is the requirements of allowing nature to take its course. If a city blocks its wind duct and blocks its river system, the air purification capacity of local atmosphere will inevitably decrease, and the water disasters (floods and drought) will of course intensify. Some Chinese cities compete to build the highest building in the world; this not only increases the risks to the building, but building structure must be reinforced, leading to more resource use; although it can increase the floor area ratio, as a whole it will utilize and waste more environment and resource capacity. If skyscrapers exceed certain height, the energy consumption for the transportation of people and materials upward and downward and for water supply to the building will increase in a nonlinear way. Once the electricity supply or some key equipments fail, the risk and vulnerability of a skyscraper will be many times higher than that of a low building. In case of fire disaster, it is very difficult to stop fire and rescue people in super high skyscrapers.

Equal Distribution of Public Resources The ecological civilization seeks harmony; one basic principle of harmony is that the elements rely on and are in proportion to each other. Under the new urbanization in ecological civilization, there should be no industry park of dozens of square kilometers that does not have any community and public service facilities and no residential area without any job opportunities within 10 km. The urban planning in the industrial civilization usually "move" people around with various investment and technologies; however, the scientific urban planning in the ecological civilization requires the integration of industry and city, work and life, the even distribution of resources, and the proportional development of various elements, so as to reducing people's needs for long-distance commuting.

Cities are the concentration centers for various social public resources, and they supply various social services to their urban residents. However, if the public resources are too concentrated, it will be hard to effectively control the scale and boundary of cities that are in monopoly position in public resource distribution; while the livability and development space for small and medium-sized cities will be squeezed.

Most of the best resources for education, health, and culture and sports in China are located in the megacities, municipalities directly administrated by the central government, and provincial capitals. Big cities keep expanding because of the high concentration and monopoly of the social public resources. For example, in Beijing, there were 91 universities or colleges in 2012, and the total number of undergraduates was 577 thousand; and 209 thousand postgraduate students were studying in 52 universities and 117 research institutions. Big cities not only concentrate the best public service resources, they also control and decide the allocation of high-quality infrastructures and economic resources. Transport hubs are usually located in the big cities and rarely in medium and small cities. For example, in the Dongdan area in Beijing, there are three national hospitals: Concord, Tongren, and Beijing Hospital. In the Zhongguancun area in Haidian District concentrate many famous universities, institutes of the Chinese Academy of Sciences, and other national research institutions of various Chinese ministries.

However, in developed countries, except for the financial service sector that is relatively concentrated, other public resources and industry distribution are usually scattered. For example, in the United Kingdom, the famous Oxford University and Cambridge University are not in London; the Cambridge Hospital is not in downtown, but there are many community general practitioner clinics in the downtown area. The California University has ten campuses located in various parts of California, instead of all concentrated in Los Angeles or San Francisco. The famous Stanford University is not in big city as well. The capital of Holland is Amsterdam, but the government, the royal family, and the Supreme Court are in Hague. In South Africa, there are three capital cities that are far away from each other: the administration capital (where the central government is located) is Pretoria, the justice capital (where the Supreme Court is located) is Bloemfontein, while the legislation capital (where the congress is located) is Cape Town.

To cure China's "megacity diseases," social public resources must be equitably distributed. Firstly, we should avoid the overconcentration of administrative agencies, good resources for education, health care, and culture and prevent the loss of economic of scale. For example, Brazil has moved its capital from coastal region to inland, and South Korea has moved its capital out of Seoul, in order to optimize the regional space distribution of high-quality public resources. Beijing Shougang Iron and Steel Company was moved out of Beijing for the purpose of optimizing Beijing's industrial structure and promoting green development. Such relocation should not only cover the manufacturing industries but also the high-quality service resources.

Secondly, it is necessary to highlight the city infrastructure's function of serving the public, not its ownership. For example, if the second capital airport is built in Tangshan City or Baoding City, the intercity fast rail makes the airport easily accessible from Beijing, which can not only significantly reduce the resource, environment, and population pressures for Beijing but also help the small cities near Beijing to adjust their economic structure and improve their environment. The rail transport of big cities can extend and connect medium and small cities around them, so that these cities can effectively share some city functions, prevent the spatial fragmentation of city system, and improve the integration of cities. The connection and integration of the hard and software facilities among neighboring cities can break up jurisdiction isolation and realize mutual support.

Thirdly, the urban space should improve the integration of industry and city, and integration of various functions, in order to avoid the long-distance commuting between people's homes and working places, the zoning of a city based on different functions, and the related waste of resources.

Fourthly, it is necessary to scientifically understand the connotation of "increase the population density in the built-up areas." In China, the concentration of population in downtown area is very prominent in every city. Almost all the big cities copy Beijing's ring road pattern. The population density in downtown area is over 20 thousand people per square kilometer. Meanwhile the population density is very low in the Industry Parks and new city districts. Therefore, even though it is necessary to "increase the population density in the built-up areas," cities also need to move people away from the old districts and downtown area. Otherwise, it will be difficult to completely solve the urban chronic diseases of traffic congestions, water shortage, high pollution, and high housing prices.

The scientific space distribution that embodies the ecological civilization ideology is not only a special characteristic of the new urbanization but also a safeguard for the new urbanization. Understanding and accommodating nature not only can reduce the resource consumption for fighting against the nature, such as long-distance water diversion or overextraction of deep groundwater, but also save environmental cost of social functioning. To implement the scientific space layout, distribution, and urban planning, the government needs to strictly follow the ecological redline threshold and the boundary for city development that have been defined based on the environmental capacities. The governments at various levels should enforce and abide by the laws, instead of violating them. This requires improving the official assessment scheme for urban development and including new criteria, such as natural resource asset and debt, ecological benefits, employment, public health, and so on, and reducing the importance of economic growth rate. With regard to specific policy instruments, the government should rely on such economic policy instruments as progressive resource consumption tax and ecological compensation mechanisms, to guide and support the equal distribution of social public resources and the sharing of infrastructure among different cities in a region, to gradually solve the overconcentration of public resources.

It is unprecedented in the human history to carry out new urbanization in China, a country of highly diverse natural resource distribution, very vulnerable ecological environment, and huge population. The new urbanization needs to meet the demand for green, livable, and human-oriented cities, keep city development within the rigid constraints from natural resources and environment, and resolve multiple enormous challenges of social and economic development. Technologies can mitigate the resources and environmental pressures at microlevel; however, it is more important to strengthen ecological civilization construction, respect and accommodate nature, build a scientific macro city distribution pattern that is in harmony with resource and environment carrying capacities, define and exert the roles of the market and the government, and equalize the distribution of social public resources, to ensure green and healthy urbanization in China.

4.5 Coordinated, Balanced, and Harmonious Development: A Case Study of the Yanjiao Town

Yanjiao is a town of Sanhe County, Hebei Province, that is 30 km from Tiananmen Square in Beijing, and its total population is half a million.[9] Driven by market forces, Yanjiao Town has achieved some degree of population and industry integra-

[9] Based on my field study in Sanhe County, Yanjiao Town, in December 2013. During that field study, I realized that the Yanjiao town case actually is a typical phenomenon in China.

tion with Beijing. However, because of the fragmentation of jurisdictions and administration, there is limited connection between Yanjiao and Beijing in such infrastructure facilities as transportation, electricity supply, and water supply and social services, for example, education, health care, and social security, which impede the effective moving of population and industries away from Beijing.

The advantage for Yanjiao Town is that it is very close to Beijing; the disadvantage is that it is not part of Beijing's jurisdiction. As a result, on the one hand, Yanjiao becomes the extension of Beijing's boundary; on the other hand, it is a city independent of Beijing. Sanhe County follows the strategy of "seamless, no discrimination, smooth connection" in receiving the outgoing population and industries from Beijing, trying to integrate itself into Beijing's half-hour economic circle. Sanhe County tries to take the initiative of connecting itself with Beijing in infrastructure construction. Transportation is one of the most important aspects for the connection. Therefore, Sanhe County welcomes the operation of Beijing's public transport in its jurisdiction and integrates its local public transport with that of Beijing. It has started the preparatory work for seeking Beijing to build a light rail from Beijing to Yanjiao Town and for building connection overpass with Beijing at the Yanjiao Exit of the Beijing–Harbin Highway. It has brought in Beijing's local grid network for electricity supply and the reclaimed water from Gaobeidian wastewater treatment plant as water supply for its power plant; for heating, Beijing Heating Company has invested over 400 million *yuan* to transform its power plant heating system to supply heating for Yanjiao.

Meanwhile, as a city independent of Beijing, Yanjiao Town has built various thematic industry parks to attract and host the transferring industries from Beijing. It uses its industry parks to make itself supplementary to Beijing in economic development. The Sanhe National Agriculture Science Park has become a base for the deployment of science and technology of modern urban agriculture and leisure, tourism, and sightseeing agricultures. The science and technology incubator park has attracted 81 high-tech incubator projects from Zhongguancun in Beijing, among which 24 projects have already graduated. The total output of Sanhe industry parks is over 70 % of Sanhe County's output; they contribute over 80 % of the county's fiscal income. Apart from the industry development, Sanhe County tries to build a city with ecological environment suitable for living and working. It invests a lot on ecology and environmental projects and follows Beijing's standard in urban environmental construction. It has implemented a long list of projects to improve local environment and protect local ecosystems. The urban green area percentage, green land coverage percentage, and public green land per capita are 43.3 %, 38.6 %, and 12.3 square meters, respectively. It has eliminated 53 production lines in 23 cement plants and mineral power enterprises, phasing out 15.57 million tonnes of production capacity, closed 30 cement blending plants, and phased out 12.6 million cubic meters of production capacity; it is taking measures to shut down mineral mines, cement, and other polluting enterprises.

The development of Yanjiao Town has contributed to share some functions of Beijing and to some extent has alleviated the resources, environment, and population in Beijing. Among the 500 thousand permanent population in Yanjiao Town

(230 thousand have local *hukou*), 150 thousand work in Beijing and more than 100 thousand work in Yanjiao Town. Yanjiao plans to increase its population to 600 thousand by 2015. Although in fact Yanjiao Town has already been the extension of Beijing's boundary, however, the institutions do not allow for this. Around 400 thousand people commute between Sanhe County and Beijing everyday; however, there is no fast rail transport, and people have to rely on buses and private cars, which cause great traffic congestion and significant time and resource losses, environmental costs, and frustration, even social disability and security pressure.

The differences in education, health care, and social security impede the population transfer from Beijing to Yanjiao Town. Even though Yanjiao Town has made great efforts in improving local health care and nursing home industry to try to attract people from Beijing to move to Yanjiao, however, due to barriers, such as different social security system, ratio and range on medical cost reimbursement, and so on, these efforts have not worked. People with Beijing *hukou* or work in Beijing don't want to move to Yanjiao Town because of jurisdiction fragmentation. Due to the restrictions on local government size and fast urbanization in Yanjiao Town, the quality and quantity of such public services as education, health care, and public security are insufficient; the education resources are unequally distributed between urban and rural areas, and there are too many students in each class in some schools; the total amount of health-care resources is insufficient, especially lacking high-level health-care professionals; the public order is not so good due to insufficient policemen and judicial officials.

Under current hierarchy administration system, Yanjiao is a town located at the bottom of the administrative ladder, while Beijing is a city at the provincial level and directly administrated by the central government. Even though that the population and economic scale of Yanjiao Town has reached the Chinese official definition for a city, however, due to the different positions of Yanjiao and Beijing in the administration hierarchy, it is hard for these two cities to make coordination dialogue. However, the specific situation of Yanjiao Town is a mirror of the overall picture. China needs to modernize its national governance system and governance capabilities through comprehensive and in-depth government reform. Whether or not the Beijing–Tianjin–Hebei City agglomeration can realize coordinated development relies on defining and regulating government roles, allowing the market to play a decisive role in resources distribution, especially in the relocation of population and industries. These reforms will have a far-reaching impact on the development of Chinese cities.

If Yanjiao Town is not the extension of Beijing's boundary, then Beijing should narrow and stick to its function definition, strictly limiting Beijing's development within its boundary. Beijing should not pursue unnecessary and unimportant functions. The core functions for Beijing include national political center, cultural center, international communication center, and science innovation center. Only in this way can Beijing be on the path of scientific, rational, high-quality, and sustainable development. For example, there are 91 universities and colleges and about one million undergraduate and postgraduate students from other parts of China, and many of the universities are among the best ones in China. The high concentration of good universities and education resources is not a core function of Beijing. On the other hand, if a convenient and fast speed intercity fast rail transit can be built between Beijing and its neighboring cities, it can effectively reduce its population and resource pressures in Beijing and realize the integrated development of the Beijing–Tianjin–Hebei City agglomeration.

Yanjiao Town, as an independent city, must have its own development positioning and function definition. In this way, Yanjiao Town will become a member of the Beijing–Tianjin–Hebei City agglomeration, rather than just boundary extension of Beijing. Thus, it is very critical to promote the integration of industry development, infrastructure construction, and social security between Beijing, Tianjin, Yanjiao, and other cities in Hebei Province. The cities around Beijing are not and should not be just the boundary extension of Beijing. They should have their own complete city functions, be independent, and meanwhile be coordinated and complementary to each other. Beijing and its neighboring cities should carry out industry division and cooperation and encourage the industry transfer from it to surrounding cities. The people who move out of Beijing should be able to continue enjoying high-quality social public services. Beijing and its surrounding cities should establish and maintain a long-term and effective cooperation relationship in environmental governance and ecological preservation.

Chapter 5
Resource Nexus and Ecological Security

The volume of ecological capacity of a region is determined by various natural resource factors. These natural resource factors are interlinked, and affect each other, some of which may have higher influences than the others and play a major binding or supporting role. They support the social and economic development of China jointly. From a long-term perspective, during the ecological civilization construction process, China should always follow the systematical nature of ecosystem, pay special attention to the nexus characteristics of resources, and minimize the cask effect of the nexus. China should define the minimum redlines for the development of its ecosystem and to accelerate its overall transformation toward an ecological civilization.

5.1 Resource Nexus

There is a Chinese saying that "pull one hair and the whole body is affected." That is to illustrate the interrelations among various components of a system and between the components and the whole system. The systematical properties of ecosystem are about the components of an ecosystem and the functions of each component, while the concept of resource nexus highlights the interrelationship among all elements and among different ecosystems.

The discussion about resource nexus mostly happens at the international political level. For example, Andrew-Speed[1] considers resource nexus mainly from the perspective of supply security, which covers three layers, including market nexus, strategic nexus, and local nexus reaction. Because of globalization and technology innovation, fluctuations in the market of one resource in one region may influence the production and supply of the same kind of resource in another region; one

[1] Philip Andrew-Speed, speech on seminar in Beijing, September 15, 2012, German Marshall Fund Transatlantic Academy.

J. Pan, *China's Environmental Governing and Ecological Civilization*,
China Insights, DOI 10.1007/978-3-662-47429-7_5

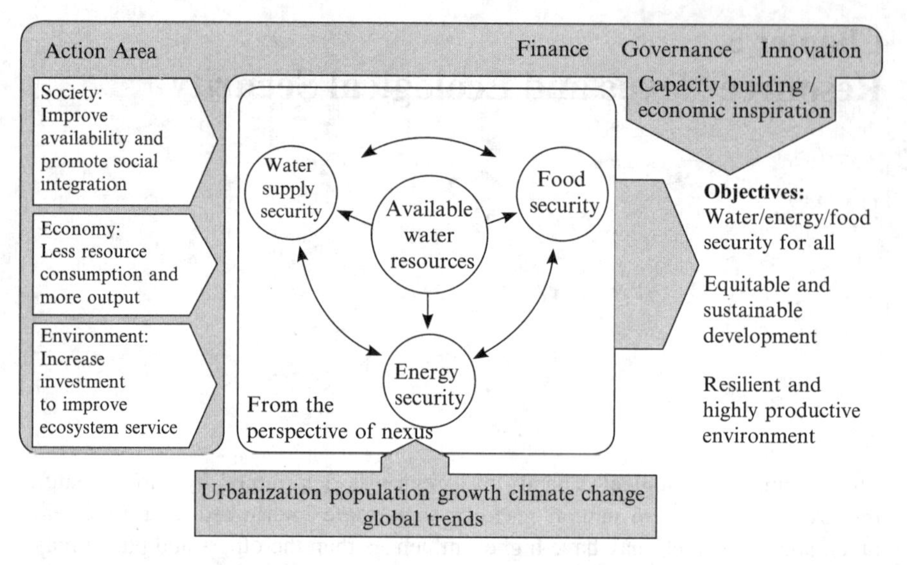

Fig. 5.1 The security implication of resource nexus

country's policy may affect the resource security of other countries and regions. Similarly, one country's policy on one kind of energy resource may affect the market situation of other kinds of energy resources in other countries. For example, if EU enacts a policy to promote the use of bioenergy, say biofuels, it may have impacts on the food supply in African countries. Also, in the case of rivers that pass multiple countries, the water resource utilization decisions and activities of upstream countries will have huge impacts on the water security of the downstream countries. The energy and water policies of different countries are usually made separately, but they can affect each other. After all, land is the foundation for fossil fuels, solar, wind, and biomass energy; while water plays a crucial role in the production and supply of energy, minerals, and food. A typical example is the important role of water use in the exploration of shale gases (Fig. 5.1).

However, more discussions about resource nexus come from the perspective of development. The Millennium Development Goals (MDGs) set by the United Nations in 2000 mainly focused on the resources related to poverty reduction and ignored the nexus among natural resources. For example, among the MDGs, water security is limited to accessing safe drinking water and clean water; other human water use and water use for ecosystem are neglected. Energy security was defined as accessing clean, reliable, and affordable energy services, including energy use for cooking, heating, lighting, communication, and production; whereas the land, water, and food issues in energy production and consumption were not considered. As for food security, which the Food and Agriculture Organization of the United Nations (FAO) defines as accessing to sufficient, safe, and nutrient food, the MDG focused on meeting people's basic need for food and food preferences for an energetic and healthy life. In other words, the MDG for reducing hunger was confined to food and did not cover its relations with other issues. During the implementation of MDGs, people found that different goals are closely interlinked. Therefore, in the United Nations (UN) process to set the "Post 2015 Development Agenda" and "Sustainable Development Goals," resource nexus has been identified as an important issue. Some think tanks have made several related studies.[2] The UN Conference on Sustainable Development held in 2012 in Rio de Janeiro decided to set up an open working group to propose the sustainable development goals. In the draft Sustainable Development Goals submitted by the open working group, in principle each goal also focuses on single factor; however, some specific goals also include some nexus considerations.[3] For example, in the specific goal for water, it clearly includes some contents on the protection and restoration of ecosystems.

The simplest description about the nexus among water, energy, land, and food is that these factors are closely interrelated, depend on each other, and constrain each other. For example, food production needs to use water, land, and energy as resource input. Grain output depends on the coexistence and joint working of these resource inputs. In the absence of land, water alone cannot support grain production; meanwhile, without water, land is desert and no crop can grow on such land, and the grain production is also impossible. Traditional agricultural production does not rely on commercial energy, and energy seems an unimportant factor for grain production. However, modern agricultural production depends on agricultural machinery, irrigation and drainage system, pesticides, and chemical fertilizers, and commercial energy consumption is indispensable. Especially nowadays, fossil fuels are depleting, and solar photovoltaic technology for power generation needs to have solar panels on large areas of land to collect sunlight; the land cannot be used simultaneously for grain production as crops cannot grow on land where the solar panels block the sunlight. Biomass production directly competes with grain production for land and water resources. Biodiesel production needs not only large size of arable land but also enormous water resources. It is estimated that to produce 1 l of biofuel, the soy beans needed consumes as much as 10 tonnes of water during their growing period through evaporation; the sugar canes needed consumes 2 tonnes of water, while the beets needed has a water consumption of 0.5 tonne.[4] When water is used for food production, the calorie output of each tonne of water is only around 3000 KJ; the market value of the food products is in many cases less than 1 US dollar[5] (see Table 5.1).

---

[2] SEI. (2011). Understanding the Nexus, background paper prepared for Bonn conference on Water, Energy and Food Security Nexus: solutions for the green economy, Nov 16–18, 2011.

[3] UN OWG (Open Working Group) on Sustainable Development Goals, July 2014.

[4] Hoogeveen, J., Faures, J.-M., and Van de Giessen, N. (2009). Increased biofuel production in the coming decade: to what extent will it affect global freshwater resources? *Irrigation and Drainage*, *58*, 158–160, Wiley.

[5] Molden, D., Oweis, T., Steduto, P., Bindraban, P., Hanjra, M. A., Kijne, J. (2010). Improving agricultural water productivity: Between optimism and caution. *Agricultural Water Management*, *97*, 528–535.

Table 5.1 Water productivity in the production of different farm products

	Wheat	Potato	Tomato	Apple
Kcal/m^3	660–4000	3000–7000	1000–4000	520–2600
\$/m^3	0.04–1.2	0.3–0.7	0.75–3.0	0.8–4.0

The nexus among resources is reflected in the mutual influences among different resources when natural situations change. In years of favorable weather, food production is high and the demand for energy in crop production is low. While in years of draught or flood, crop production needs to consume larger quantity of modern energy service for irrigation or drainage. Such linkage means that, once the security of one resource is at risk, it will affect the security of other resources. For example, when water security is at risk, water shortage leads to drought, and land resources are of little output, this will further threaten land security and food security. Below is a closer examination of the resource nexus issue in China. Due to its current development stage and population profile, China faces challenges in maintaining the security of multiple resources. Therefore, interrelated resource security is a kind of nexus security among each component resource and the whole resource system, instead of the security of each single resource. That is because China's per capita average ownership of ecosystem or natural resource is quite low, which leads to a quantitative constraint on the country's development. Moreover, quality of natural resources is another problem. Environmental pollution, water quality degradation, and severe heavy metal pollution on soil have already happened and will probably take 30–50 years to control the pollution and clean up the environment. The environmental pollution also leads to food security problems. Grains produced from the land polluted with heavy metal also have high heavy metal contents and are not safe to use as food. Increases in such grain output cannot contribute to the improvement in food supply security. Hence, quality of resources is also crucial to food security.

Nexus security, or resource nexus security, means that countries have to choose between different needs. To safeguard energy security, a country can grow crops for biodiesel at large industrial and commercial scale, which is totally feasibly under current technology level and capacity. Meanwhile, safeguarding energy security in this way will lead to higher food security risks as more land used for biofuel production leads to less land for food production. Similarly countries need to balance and choose between ecological security, energy security, and food security.

Coal mining can contribute to energy security, but such activities could damage the underground water system, which has been formed through many years, and its restoration may take 30–100 years. Under such circumstances, the whole ecosystem and food production will be severely constrained or disrupted due to underground water system damage or pollution. In some parts of China, such things have already happened. Inappropriate mining practices and exhaustive mining have caused damages to underground water system below some goals and negative impacts to ecosystem and food security.

5.1.1 Global Significance of China's Resource Nexus Security

Due to resource nexus, the stock and variability of China's ecological resources has global influence. First of all, the Chinese economy highly depends on foreign trade. Many raw materials, such as petroleum, iron ore, and some main agricultural products, are of very high import dependency. Since 2010, over 60 % of China's petroleum supply has been depending on import. At the same time, China's annual import of iron ores has reached 800 million tonnes, and most of its soybeans for cooking oil production are imported. China's high dependency on the international market and its huge imports of resource products provide some demand and business opportunity to the global resource markets. Large resource import can to some extent help support the balancing Chinese ecosystems, but it also increases the vulnerability and risks to China's resource security.

For example, China's import of soybeans in 2011 was 4 % less than in 2010; however, due to price increase, the amount it paid for the import was almost 20 % high than the year before. China imported 260 million tonnes crude oil in 2011, 6 % more than in 2010; meanwhile, the costs of the import jumped 45.3 %. China has little control on the prices of its imports. The United States can to a large extent control the crude oil price on the international market; China still lacks such power and is merely a price taker. This is most obvious on the international crude oil market; the same is also true for the international grain market. China's high dependency on the international market is of great significance to its nexus security of natural resources.

China's resource security to some extent depends on the global market. Climate security is a serious problem in China. The geographic distribution of climate capacity determines the distribution of Chinese population and economy. Due to historical climate evolution and ongoing global warming, the climate capacity in Northwest China is continuously shrinking, and climate immigrants are inevitable. In Shaanxi Province, the local government made a 10-year plan in 2011 to resettle as many as 2.4 million people. Obviously, this is because the climate capacity in some areas was insufficient to support their population size, and some of the local people had to be moved to other areas. Climate immigration has also been going on in Ningxia for over a decade. Nowadays, extreme climate events, such as floods and draughts, and sea level rises directly influence the areas most advanced in economic development in China, including the Pearl River Delta, the Yangtze Delta, and the area around the Bohai Sea. It goes without saying that the low-lying coastal areas are also subject to severe climate change impacts.

Global climate security issues induce a series of interest and security conflicts among nations. International trade disputes are easily raised because of climate change policy differences among countries. Countries with stricter regulations on GHGs emission reduction may worry that their products lose comparative advantages in the market to other countries and try to avoid the comparative advantage change through levying border tax. For example, the EU aviation carbon tax is a kind of border tax, and it leads to some trade disputes between the EU and other countries.

Climate change policies can also lead to concerns about energy security. Climate change may induce security problems at both national level and international level, such as climate immigrants and climate refugees. If climate refugees appear in developing countries, there are surely sources of security challenges to the world. The security of water, food, species, and energy is closely connected; security problems in one single element in one country, through nexus transfer, may lead to an overall global security issue.

Long-distance transportation of bulk resource products, including grain, petroleum, and ores, could raise concerns about security during transportation. Since 2010, China's petroleum import has been over 300 million tonnes per year, mainly through sea transportation. Meanwhile, aviation and marine transportation is accompanied with high security risks. Pirate activities, international disputes about maritime rights and interests, and regional military conflicts are all causes of severe security risks to the international transportation of bulk resource.

China should include global resource nexus security in its national security agenda, seeking for international cooperation and taking responsibility accordingly. First of all, at the global level, an international setup for international governance is urgently needed. International cooperation is beyond the capability of any single country. Nexus security is a global issue, which needs multiple and bilateral cooperation to set up an international security system. As a responsible large developing country, China takes climate change, international trade, and eco-security issues very seriously. It is not only a passive participant in the global process but also actively contributes to the negotiations and dialogues on how to reflect and protect the resource nexus security through international legislation.

Apart from cooperation at global level, regional, multilateral, and bilateral cooperation is also indispensable. Moreover, regional and bilateral dialogues, communications, and cooperation are crucial to the security of resource nexus. Particularly, China's further development relies on both international and domestic markets and resources from both China and abroad, where a proper coping strategy for international security is necessary. The international political game for resource nexus security is actually about protecting national economic benefits. Therefore, economic discourse power is very important. China is the largest importer of iron ores, but it has little influence on the iron ore prices on the world market. Chinese companies, especially state-owned companies, should pay high attention on their capacity building to promote the country's resource nexus-related economic security.

When it comes to the domestic responses, quality of resources is crucial. China needs to enhance its security administration on the quality of its resources and monitor and control invasive exotic species and pollutants emission. Heavy metal pollution of the soil also needs to be controlled and included in the government plan as soon as possible.

5.2 Identification of the Ecological Function

China has its unique landscape and climate characteristics, which shape its special natural ecosystems. China's topography is high on the west and low on the east, and most of its rivers flow from the northwestern plateau toward the seas in the southeast. China has a monsoon climate, which makes its precipitation unevenly distributed temporally and spatially. In summer, the rainfalls are intensive and the typhoons are frequent; while in winter, cold air from the north causes low temperature and low precipitation. The northwestern plateau and mountain areas have a weak ecosystem with low carrying capacity, but they are important natural ecological shields for the agricultural and industrial production, and urban and human settlement environmental in the southeast. Protecting the fragile ecosystem of northwest China can benefit the interrelated national ecosystem. To respect and follow the law of nature, China needs to take into account the interconnections among its ecosystems in different regions in its development and utilization of natural resources (Fig. 5.2).

Currently China is in the process of rapid industrialization and urbanization, some industries are moving from coastal areas to inland areas, while most of the inland areas are experiencing large-scale urbanization. Meanwhile, the fragile ecosystems and environments in the middle and the west are under threat. It is urgently needed to make national overall plans for population distribution, economic layout, land use, and urbanization work. The main functions of each region urgently need to be defined based on their environmental carrying capacity and their current developing intensity and remaining potential. This can help clarify the development directions, improve development policies, control the development intensity, and

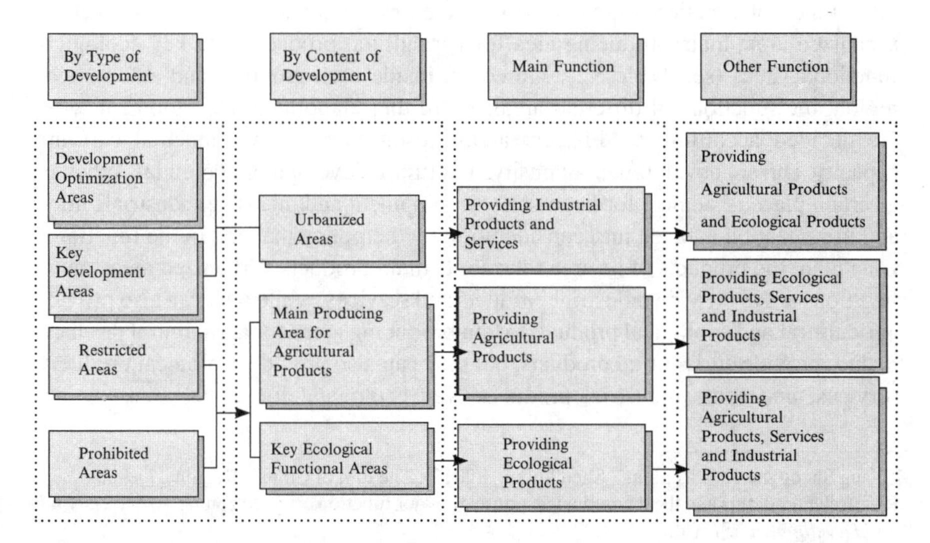

Fig. 5.2 China' main functional areas and their functions (Source: State Council of China (2010). National Plan on Main Functional Areas)

regulate the development activities, in order to make sure that the population density and economic activity intensity are in line with local resources and environmental conditions.

The main function of an area is the principal function or primary function among the multiple potential functions of the area's natural resources from the perspective of ecosystem production. An area's main function can be decided based on the goods it can provide, or whether it is in the position to provide industrial products and services, or agricultural products, or ecological products and services. If a place is crucial to the national overall ecosystem security, its main function should be providing ecological and agricultural products and services, and providing industrial products is a secondary function of this place. Otherwise, its productivity of ecological products will be damaged. For example, the main function of grassland is to provide ecological products; if it is overgrazed, grassland degradation or desertification will occur. The main function of areas with good condition for farming shall be providing agricultural products. If large area of arable land in such areas is acquired for other uses, the productivity of agricultural products will be damaged. Therefore, different areas' main functions should be differentiated, and their main contents and major tasks of development are decided accordingly. From this perspective, defining the main functions of different areas is a way to respect and follow natural laws and an effective measure for stimulating the development paradigm transformation at local level.[6] China began to plan the main functions of different regions in order to protect its local resources in 2007.[7]

In the National Plan for Main Functional Areas, China divided different provinces, autonomous regions, and municipalities into four kinds of functional zones based on their development types, which are development optimization zones, key development zones, restricted zones, and prohibited zones. Besides, there are three kinds of functional areas based on their main economic activities, such as urbanized area, main producing area for agricultural products, and key ecological functional areas (see Table 5.2). Moreover, inside each province and autonomous region, the functions of different areas are further elaborated. Development types are decided according to different areas' resource and environmental carrying capacity, current development intensity, and future development potential. Whether a certain place is suitable for large-scale development and intensive industrialization and urbanization is taken into consideration by the policy makers. In the functional zone plan, the products of an area refer to its main products. Urbanized areas' main functions are to provide industrial products and services, while they can also provide agricultural and ecological products. Main producing areas for agricultural products mainly provide agricultural products, but they can also provide ecological products, services, and some industrial products. Key ecological functional areas mainly

[6] Wang Shengyun., Ma Renfeng., Shen Yufang. (2012). The turn of China's Regional Development Research Paradigms and the theoretical response of major function oriented zoning. *Areal Research and Development, 31*, (6).

[7] Opinion of the State Council on the Preparation of the National Main Functional Area Planning (State Council (2007) No. 21).

Table 5.2 Indicators for national plan on land spatial development

Indicators	2008	2020	Changes	
Intensity of development (%)	3.48	3.91	0.43	12.36 %
Urban area (1000 Km^2)	82.1	106.5	24.4	29.72 %
Rural area (1000 Km^2)	165.3	160	−5.3	−3.21 %
Arable land (1000 Km^2)	1217.2	1203.3	−13.9	−1.14 %
Forest area (1000 Km^2)	3037.8	3120	82.2	2.71 %
Forest coverage (%)	20.36	23	2.64	12.97 %

Source: Calculated based on the National Plan for Main Functional Areas

provide ecological products and service, while in some cases, they also provide some agricultural products, services, and industrial products.

According to the National Plan for Main Functional Areas, land use for human settlement in China should be kept below 3.91 % of the national land area. In particular, urban land use should be at maximum 106,500 km^2, land for rural human settlement should decline to less than 160,000 km^2, and the arable land use for new construction areas should be kept below 30,000 km^2. The minimum area of arable land possession should be 1,203,300 km^2 (1.805 billion mu), among which basic farmland should be at least 1,040,000 km^2 (1.56 billion mu). The forest should increase to 3.12 million square kilometers, and forest coverage should rise to 23 %, and forest growing stock should reach 15 billion cubic meters. Grassland proportion in total land use is above 40 %. As indicated above, with the expansion of urbanization, land use for rural human settlement and farming will both decrease, as part of them will change into urban and industrial land (see Table 5.2).

Western areas and mountain areas are ecological shield for the rest of the country and have mostly been categorized as prohibited zones for development. In these regions, many areas are earmarked for natural and cultural resources protection, including national natural protection zones, world cultural and natural heritage, national scenic areas, national forest parks, and national geological parks. Provincial prohibited zones include all kinds of provincial or local natural cultural resource protection zones, key water source areas, and other prohibited areas for development decided by provincial governments (Fig. 5.3).

5.3 Ecological Restoration of Cropland

Ecological restoration of cropland is an important administrative measure taken by the Chinese government since 1999, in order to stop large-scale land degradation because of improper land use. It includes converting cropland to forest, grassland, and water. Converting cropland to forest is mainly about the cropland in mountain areas, where the cropland had been expanded through slashing and burning forest and woods, especially in places where the slope is steeper than 25°. Besides, this program also covers the cropland obtained through destroying forest in semi-wet

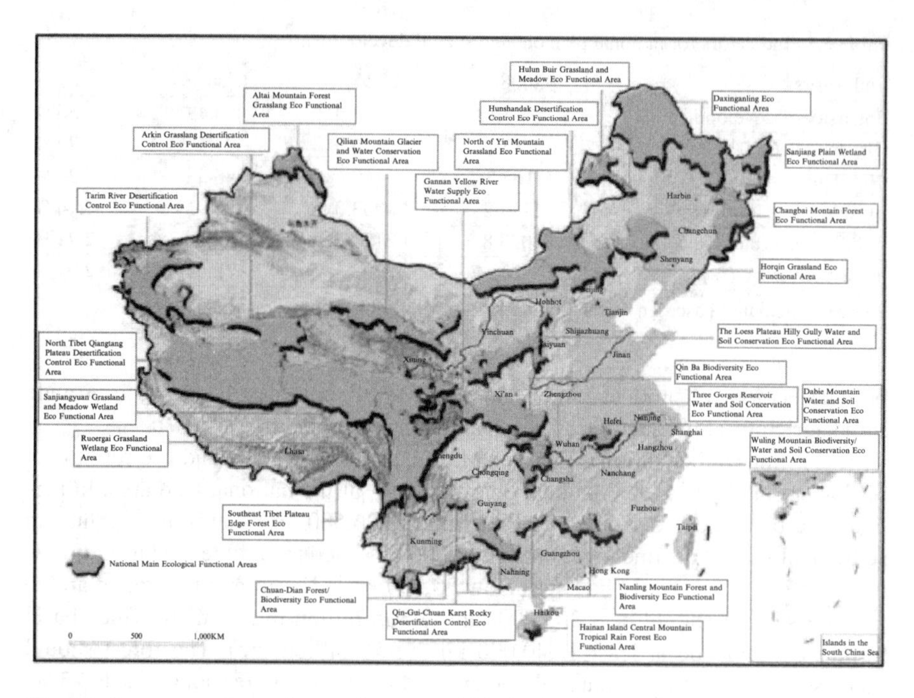

Fig. 5.3 Key eco service regions with areas prohibit for development (Source: National Plan for Main Functional Areas, 2011)

ecological vulnerable areas. For decades, China had taken grain production as the most important thing in its socioeconomic development and encouraged people to massively destroy forest, wetland, and grassland for more arable land, which caused severe imbalance of natural ecosystems. The converting cropland to forest program began in 1999 and has been the ecological construction program with strong policy intervention, the highest cost, and the widest coverage, and the highest level of public participation. It is also a huge program that focuses on benefiting farmers. Investment from the central government alone exceeds 430 billion yuan and has been the biggest ecological construction project in the world. The Chinese government has issued multiple guidance and instructions on the program, including *Instructions of the State Council on Further Pilot Work For Returning Cropland to Forest and Grassland* (State Council [2000] No. 24) and *Instructions of the State Council on Further Improving the Policy of Returning Cropland to Forest* (State Council [2002] No. 10), and set the target of returning 14.67 million hectares of cropland to forest during the 2001–2010 period. The National Ordinance on Returning Cropland to Forest was also enacted. The national forestry authorities conducted in-depth surveys and held consultations with relevant local authorities from provinces, municipalities and autonomous regions, and relevant experts. On the basis of national regulations, target, and investigation and consultation results, the China State Forest Administration, jointly with the National Development

Reform Committee, the Ministry of Finance, the State Council West Development Office, and National Administration of Grain, made the *National Plan for the Program of Returning Cropland to Forest* (2001–2010). The objectives of this plan are (1) to return 14.67 million hectare of cropland to forest; (2) to grow 17.33 million hectare of forest on barren hills and wasteland, both of which included the pilot activities for returning cropland to forest during 1999 and 2000; (3) to fundamentally realize the target of returning cropland to forest on steep slopes; (4) to generally improve the conditions of seriously desertified arable land; (5) to increase the forest and grass coverage in the program areas by 4.5 percentage points; (6) and to significantly improve the ecological conditions of program areas.

From 1999 to 2004, the central government allocated among local governments the tasks to return 19.1655 million hectares of cropland to forest, which included returning 7.8862 million hectares cropland to forest and growing 11.2793 million hectares of forest on barren hills and wasteland. So far, all the provinces and autonomous regions have generally completed their tasks, and some provinces and autonomous regions have even surpassed their targets. The inspection results indicate that the quality of work under the program is good.

From 2000 to 2004, the central government accumulatively provided 74.803 billion yuan of subsidies for the implementation of this policy, which includes 14.374 billion yuan as subsidies for seeds and saplings, 121 million yuan as expenses for preparation work, 6.285 billion yuan of living allowance, and 54.023 billion of grain subsidies. Implementation of the returning cropland to forest policy leads to an annual forest area increase of 6.67 million hectare during 2002–2004, compared to a former annual increase of 4–5 million hectares. The forest area increase due to the program of returning cropland to forest respectively accounted for 58, 68, and 54 % in the total national afforestation area in 2002, 2003, and 2004, and in some of the western provinces, this share exceeded 90 %. Implementation of this program has improved the relationship between human and the nature and changed the traditional practice of highlighting production area and neglecting productivity among Chinese farmers. Thanks to this project, soil erosion and desertification have been declining quickly and the ecological conditions in areas covered by the program have dramatically improved. According to monitoring reports by the Yangtze River Water Resources Commission, annual sediment carried by water flow decreased 80 % at the Yichang Station in the upstream of Yangtze River in 2003. The sediment carried in water flow became lower than the multiyear average in the main branches, and average sediment concentration at each station lower than Cuntan Station had declined by 50–79 %. Implementation of returning cropland to forest is the main reason behind of the declines in sediment content of the Yangtze River. In Sichuan Province, the accumulated area of returning cropland to forest reached 805.3 thousand hectares during the period from 1999 to 2004, which leads to a decrease in soil erosion by 267 million tonnes; the annual decrease was 53 million tonnes, almost a quarter of the annual total sediment left in the province. Besides, sediment concentration per cubic meter in the Min River and the Fu River, two tributaries of Yangtze River, separately decreased by 60 and 80 %. The program of returning cropland to forest has made an important contribution to push China's ecological construction to a new stage, during which damaging activities and improving efforts are in a stalemate.

Ecological construction activities similar to returning farmland to forest have also taken place in some developed economies. The American government passed the Food Security Act in 1985 and carried out the Conservation Reserve Program (CRP). CRP is based on voluntary participation and government subsidies, which encourages farmers to leave farmland fallow for tree and grass growth for 10–15 years, so as to restore vegetation on the fallow land. CRP focuses on farmland that has soil erosion problems and vulnerable to environmental changes and aims at improving water quality, controlling soil erosion, and improving biodiversity. The amount of subsidies is proposed by farmers and evaluated by experts hired by the government, and the subsidy includes a land rent of 44 USD per acre and 50 % of the costs for planting trees and grass. As of 2002, the CRP had covered 13.6 million hectare of cropland, 55 % of which had joined the second period of CRP.[8]

The European Community started a fallow land program in 1988, aiming to reduce the huge and expensive agricultural product supply surplus and to decrease damage to agricultural ecosystem and wild animals and plants.[9] In 1992, the fallow land program became a mandatory requirement,[10] which demanded 15 % of the cropland to lie fallow. This proportion decreased to 10 % in 1996, which was equivalent to about 9.4 million acres of farmland. The fallow program improved the soil chemical contents of cropland and contributed to greater biodiversity. The European Union terminated this program in 2007 and has stopped providing subsidies for fallow land since 2008, in order to increase grain supply, to curb food price increase, and to increase biofuel production.[11]

Although United States, EU, and China have all implemented some programs of "converting farmland to ecological land use," the preconditions, targets, and impacts of their programs are quite different. First of all, their preconditions of farmland covered by the programs are different. All of the farmland to be left fallow in the EU and the United States, though it may not be of the best quality, is suitable for agricultural production. In contrast, in China the farmland to be returned for ecological land use is on slopes steeper than 25°, arid grassland, and wetland. The land on steep slopes and grassland is of low productivity and rapid soil degradation and unsuitable for farming use. These extramarginal land resources were used as farmland in China

---

[8] Xiang Qing., Yin Runsheng. (2006). Review of America Conservation Reserve Program. *Forestry Economics*, (1), 73–78.

[9] "Commission Regulation (EEC) No 1272/88 of April 29, 1988 laying down detailed rules for applying the set-aside incentive scheme for arable land." EUR-Lex. European Commission. April 29, 1988.

[10] Set aside, Dinan, Desmond (Feb 20, 2014). *Origins and Evolution of the European Union*. OUP Oxford. p. 210. ISBN 978–0,199,570,829.

[11] According to one report in British Daily Telegram, during 2006–2007, food price was doubled. The price of wheat per ton reached the level of 200 pounds. And that led to the price increase of other foods, like bread. Waterfield, Bruno (September 27, 2007). "Set-aside subsidy halted to cut grain prices." *The Daily Telegraph* (London).

because of grain shortage and scarcity of arable land. Therefore, the fallow land programs in the United States and EU, are actually "rotatory rest" of arable land; the fallow land functions as farmland reserve and can be put back into cropland use whenever necessary. While in China, although the cropland returned to forest and grass could see some improvement in its productivity, due to topographical and climate conditions, it will still be unsuitable for cropland use. Secondly, the primary objectives of the fallow programs in the United States and EU are to reduce agricultural product oversupply and to reduce grain production subsidies. Environmental protection is just a secondary objective. That is also why EU has stopped its fallow program, and in the United States, the second period of CRP only lasts for 5 years, instead of 10–15 years of the first period. While in China, things are different. Arable land insufficiency and grain shortage have always existed in China, and there is no agricultural product oversupply. Moreover, before the reform and opening up, China not only provided no subsidies to agricultural products but also required agricultural products to subsidize industrial products through the price scissors.

Thirdly, the fallow programs in United States and EU are voluntary and market based. Farmers voluntarily participate in the fallow program, and the government subsidies can generally cover their opportunity costs from leaving their cropland fallow. In China, such programs are mandatory and based on government plans. In the United States and EU, farmers can decide to withdraw from the fallow programs at any time, while in China, once the farmers give up their cropland for converting to ecological uses, they are not allowed to leave the system freely. Besides, in the United States and EU, the subsidies are monetary, while in China, the subsidies are grains to meet the farmers' needs for food and some monetary subsidy for covering the farmers' living costs.

Fourthly, the fallow program in the EU only let the land to be left fallow rather than to change the land use. They did not offer subsidies to encourage the growing of trees and grass on the fallow land. And in the United States, farmers were encouraged to grow trees and grass on the fallow land, but the subsidy only covered 50 % of the cost.

However, in China, the related land is not just left fallow; instead, it involves land use change and the growth and maintenance of trees and grasses.

Fifthly, and the biggest difference is about the long-term benefits of the programs. In China, as the environmental conditions of the targeted areas are rather weak, once the cropland is converted back into forest, the benefits of ecological restoration and environmental improvement are far higher than that of the fallow programs in the United States and the EU. Besides, in China, due to the actual land use change, once cropland is converted into forest and grassland, it will be more difficult and costly to change it back into cropland.

5.4 Wooden Barrel Effect

Wooden barrel effect is also known as the wooden barrel theory or the short board effect. It refers to that, to be able to hold a full barrel of water, all boards of a wooden barrel have to be of the same length and without damage and holes. Otherwise, if one of the boards is shorter than the rest or has a hole in the bottom, it will be impossible for the wooden barrel to hold a full barrel of water. That is to say, how much water a barrel could contain is not decided by the longest board but by the shortest one.

As resources are in nexus, if the wooden barrel effects also exist in resource nexus, then what is the short board of China's natural resources? Is it land, energy, or water? Land is a spatial concept, and the value of land is subject to the influences of water and energy resource conditions. From the perspective of food production and demand, food shortage has long been a constant concern in China. China uses 7 % of world's arable land to feed 22 % of global population. It is an accomplishment, but more of a hard reality the country has to live with. China's arable land shortage is mainly because of water shortage. If the arid and semiarid areas, deserts, and semideserts had enough water supply, then China would not be so short of arable land. Therefore, water is the biggest constraint of the resource nexus. Energy security is about the quality of life. Without energy, modern living style could not sustain. While without water, survival could not sustain. In that sense, water security is linked to survival security, while energy security is the security of quality life. Some people may say that with enough energy, even sea water could be desalinated to supply fresh water. This solution won't work because it is impossible to get so much energy resources to produce enough freshwater supply for human production and living activities. Therefore, water resource security should be regarded as the short board for China's resource nexus security, because of the low per capita water resource possession, the unevenness in the temporal and spatial distribution of water resources, serious water pollution, frequent floods and drought disasters, as well as heavy losses from these disasters each year.

China is still a developing country, and its per capita average possessions of most strategic resources are very low in the world. In the recent three decades, the Chinese economy has experienced a golden period of rapid growth. The per capita income of Chinese people ranked the 110th in the world in 2012, much higher than its rank of the 175th in 1978. However, China's gross domestic production (GDP) is the second biggest in the world in 2012; the country's low per capita income rank and its high total GDP rank are incomparable and a mismatch (see Table 5.3). According to the definition by the Food and Agriculture Organization of the United Nations, per capita average food possession of no less than 400 kg is one key criterion for food security. In China, per capita food production had just surpassed this threshold in 2010. As the food consumption structure changes, China's per capita food

Table 5.3 China's GDP and per capita GNI ranking in the world

Indicators	1978	1990	2000	2009	2010	2012
GDP	10	11	6	3	2	2
Per capita GNI	175(188)	178(200)	141(207)	124(213)	120(215)	110(213)

Note: Numbers in the brackets are the total number of countries and regions in the ranking list
Source: Data of 1989–2010 comes from China National Statistics Bureau and China International Statistics Year Book 2012; while data of 2011 and 2012 comes from the World Bank database: http://data.worldbank.org.cn/indicator/NY.GNP.PCAP.CD

consumption has become lower than its per capita food production. Statistics indicate that China's oil import dependency has exceeded 50 % since 2010, while its proved and expected oil reserves are rather limited. Besides, the per capita water and arable land possessions are also very low. China's future development is constrained by its development stage and huge population base, while the rigid constraints of resource availability pose the most important risk to its nexus security.

Water resource security and land security are the preconditions for food security, while mineral resource and energy security can have externality influences on water security. Therefore, water is the key strategic resource to nexus security. China's per capita water resource possession is low, and its water supply is insufficient and unevenly distributed throughout the year and among different regions. Moreover, the existing water resources are badly polluted and the drinking water sources are under threat. Each year floods and droughts cause heavy social and economic losses and enormous environmental damages.

According to statistics on global precipitation distribution (see Table 5.4), China's annual precipitation is 25 % lower than the world average. Due to its large

Table 5.4 Global precipitation level[a]

Nations and regions	Precipitation mm	Water resource from precipitation (trillion m^3)	Population million	Area (10,000 km^2)	IIRPer capita precipitation possession m^3/person
Global	813	108.83	6969.7	133,790	15,614.8
Asia	827	26.83	4213.3	32,420	6366.9
South Asia	1062	4.76	1621.3	4480	2932.8
East Asia	634	7.45	1580.6	11,760	4715.8
North America	637	13.87	462.2	21,780	30,004.7
South America	1596	28.27	396.4	17,710	71299.4
Europe	577	13.27	740.4	23,010	17,920.3
Africa	678	20.36	1044.3	30,050	19,496.2
Oceania	586	4.73	29.3	8070	161,497.3
China	626	6.01	1384.7	9600	4342.9
Taiwan, China	2429	0.09	23.4	36	3722.1
United States	715	7.03	315.8	9830	22,261.6
Russia	460	7.87	142.7	17,100	55,114.5
Canada	537	5.36	34.7	9980	154,635.9
India	1083	3.56	1258.4	3290	2829.1
Japan	1668	0.63	126.4	380	4986.0
Brazil	1782	15.17	198.4	8510	76,496.9

[a]Data of China is cited from China Metrological Administration, and data of other regions is cited from FAO. 2014. AQUASTAT database, Food and Agriculture Organization of the United Nations. http://www.fao.org/nr/water/aquastat/data/query/index.html?lang=en

population, China's per capita precipitation possession is only one fourth of the global average, one fifth of the average in Africa, and about 2.5 % of that in Australia. Again, water resource is the short board for China's resource nexus security.

Although no one can deny the importance of water resources, the situation of high exploration rate and low utilization rate of water has not been improved. As extreme weather events frequently occur, the short board effects of water resource will become increasingly severe. On July 21, 2012, a heavy storm in Beijing caused damages over an area of 16,000 km^2, 77 causalities, and an economic loss of 6.1 billion Yuan.

The short board effects of water security are manifested in four aspects. The first one is total quantity shortage. The second aspect is imbalanced distribution among different regions; Northwest China and North China suffer serious water shortage. The third one is imbalanced distribution around the year; the precipitation in spring and winter is very low, causing widespread drought; while the high precipitation in summer often causes floods. The forth aspect is water pollution.

A fundamental principle of ecological construction is to follow the law of nature. Total volume shortage and the imbalanced temporal and spatial distribution of water resources are characteristics of China's water resource endowment, and people should respect and adapt to them. China has thousands of years of history in water management and utilization. In history, floods and droughts in the middle and lower reaches of the Yellow River used to be frequent and unpredictable. The middle stream Yangtze River, as main national grain-producing region, used to be plagued by frequent floods. Since 1949, the management of big rivers has been a top priority on the government work agenda, and lots of efforts have been made to reduce disasters related to big rivers. Channels and river beds were dredged to avoid water clogging, reservoirs were built to store water, and dams were constructed to control water flow. Thanks to these efforts, the Yellow River flood has been stopped and the regions at its middle reaches have become the country's bread basket. Floods in the Yangtze River have also been put under control. There are many different opinions regarding the ecological impacts of the Three Gorges Dam. Many of the skeptics are those who have never experienced the flood disasters in the middle and lower reaches of the River and do not understand the flood control benefits of the huge dam. The author's hometown is in a plain area along the Yangtze River below the Three Gorges Dam. Due to frequent floods in this area, the villagers used to build their houses on high soil foundations to cope with the floods. Every household had a small boat for transportation in case of floods. In winter when the farmers did not need to work in the field, they used to make large grass sacks, into which they could put soil in summer and use them for flood control. In summer, all male labors had to patrol on the dams around the clock to watch out flood disasters.

Big floods often turned the region into a world of waters and caused complete loss of harvest. In those conditions, it was impossible for the local people to develop industry and build cities. In 1954, in Wuhan City alone, over 30,000 people lost their lives due to floods. In 1998, to fight against a major flood in Jing Jiang section of the Yangtze River, the Chinese government mobilized nationwide resources, and the flooding of large cities were avoided at huge economic cost. Construction work on the Three Gorges Dam was completed in 2006, and thanks to the huge dam, people living along the Jing Jiang Section of the Yangtze River no longer need to worry about flood risks. Due to the geographic landscape and precipitation distribution along the Yangtze River, the Three Gorges Dam could effectively reduce the flood risks of the middle and lower reaches of the Yangtze River. The dam was built in a place of high mountains and deep valleys. It is a fruit of the advanced technologies of industrial civilization; but its environmental effects contribute to the transformation toward ecological civilization. The reservoir areas of the Three Gorges Dam are densely populated and high and steep mountains. The local people's livelihood depends on planting crops on steep slopes and damaging forest for firewood, leading to severe soil erosion. Hydropower from the Three Gorges Dam meets the local people's energy needs for production and life, substitutes firewood use, and brings about significant effect on ecosystem restoration.

In the Chinese history, most wars and famines were related to extreme weather events, especially serious droughts. To solve the problem of water security, irrigation was a key solution. The Dujiangyan irrigation system in Chengdu built in 256 BC was a landmark historical project of following the law of nature and making use of natural resources. As China had a high population density, droughts used to affect large numbers of people, the focuses of investment and construction for water conservation were flood prevention and irrigation system.

By 1978, there were 85,000 reservoirs of various sizes in China. Among them 311 were large reservoirs with more than 100 million cubic meter storage capacity each. 2205 were middle-sized reservoirs whose storage capacity was between 10 million and 100 million cubic meters. The other 8200 were small reservoirs with storage capacity between 100 thousand and 10 million cubic meters each. The total storage of these reservoirs exceeded 400 billion cubic meters. With the acceleration of industrialization and urbanization, the water demand for agricultural irrigation and urban industrial process has increased dramatically, and the number of large and middle-sized reservoirs has been growing rapidly. As of 2011, the number of large reservoirs had reached 567, with a total storage capacity of 560.2 billion cubic meters. The number of middle-sized reservoirs had increased to 3346, with a total storage capacity of 95.4 billion cubic meters. These reservoirs effectively provide water supply for food production, urban life, and industrial process. In 1978, the total length of embankments along big rivers was 165,000 km; by 2001, this number had reached to 300,000 km, which provides protection for 42.625 million hectare of cropland and 572 million people.

China is a big emerging economy. Its development direction and speed will have huge impacts on global economy, politics, and environment. China's successful governance experiences and good practices in safeguarding resource security offer good examples for the resource security government in other countries, especially developing countries. More than 70 % of Chinese cities and more than 50 % of Chinese population are in areas subject to frequent weather disasters, earthquakes, and oceanic disasters. More than 80 % of the poor people in China live in the ecologically sensitive, water-scarce areas. And most of the poor areas locate in places subject to important climate change impacts. Due to droughts and water scarcity, people had to leave their hometown and move to other regions. Such climate immigrations are numerous in Chinese history. Even today, despite the advanced industrial technologies, this kind of immigrations continues. In Shaanxi and Ningxia in Northwest China, climate immigration is included in local government work plans.[12] Large-scale climate immigration also takes place in Bangladesh. Because of this, many big racial conflicts have broken out between India and Bangladesh. In order to prevent the inflow of climate immigrants from Bangladesh, India has built a 4000 km barrier between the two countries, the longest such barrier in the world. China's successful experiences on the management of climate immigration could offer good reference on how to solve the immigration-related conflicts between India and Bangladesh.

5.5 Ecological Security

The 18th National Congress of Chinese Communist Party (CPC) put forward the strategy of constructing ecological civilization and the framework of ecological security. The 3rd Plenary Session of the 18th CPC National Congress decided to further enhance the construction of ecological civilization institution and to define the ecological redline threshold. Obviously, ecological security is one important component for national security system and a cornerstone of national security. It needs to be included in the strategic plan for national security. However, what is ecological security? How to better interpret ecological security? What kind of ecological security is needed? And how should China realize or establish ecological security?

Ecology refers to the dynamic equilibrium relationship between human and the nature. Security is the elimination or control of the actual or potential threats to national territory, economy, social development, and social stability. Therefore, ecological security refers to avoiding or effectively controlling damages to the harmonious relationship between human and the nature because of the negative impacts of nature, human activity, and market-related factors and the consequent actual or potential threats to national sovereignty, stability of the country, national economy, and social development.

The relationship between human and the nature is not equal. Human beings are secondary to the nature, as all materials for the existence, development, and well-being of human society come from the nature. The status and changes of the natural environment have clear security implications to human socioeconomic units, which are sovereign states. As the objective nature and diversity of the environment, ecological security can be defined in broad sense and narrow sense. Broadly speaking, ecological security is a status, under which the status and changes of ecosystem and natural environment do not pose actual or potential threat to national security, or such kind of threats could be effectively avoided or controlled. That means, when human and the nature are harmonious, countries have ecological security; while when the harmony between human and the nature is broken due to any reason, countries' ecological security is in danger. As the ecosystem and natural environment consist of various elements, the aggregation security of these elements is ecological

[12] Pan Jiahua., Zheng Yan., Bo Xu. (2001). Raise the alarm: Climate immigration. *World Affairs*, (9).

security. Therefore, if any element is in danger, countries lose ecological security. Single element's security is ecological security at element level, which includes the following contents:

Water security: Water is the source of life, and the quality, quantity, and temporal and spatial distribution of water are the essential factors of water security.

Energy security: Energy security mainly refers to the supply security of fossil fuels, especially petroleum supply security. Because of their low price and high heat value, fossil energy is the energy power foundation for socioeconomic development of a country. Lack of fossil fuel supply security inevitably puts national security under threat.

Mineral resource security: Many rare mineral resources have a security implication to national defense and economic development. Their production and supply risks are threats to national security.

Environmental security: Biodiversity and air pollution, and so on.

Food security: Food is the output of natural system. It is indispensable to people's survival. Dramatic declines in food production or insecure food supply, obviously, have great impacts on national security.

Climate security: Climate system changes, especially global warming, will threaten natural and socioeconomic system.

Ecosystem security: Soil erosion and droughts cause desertification and ecosystem degradation, which harms the harmonious relationship between human and the nature and increases ecosystem vulnerability and instability.

Ecological security in broad sense covers all the factors mentioned above, while ecological security in narrow sense mainly refers to ecosystem security; in some cases, climate security is also included.

The characteristics of ecological security include the following aspects.

Nature of nexus: Ecological security consists of the security of multiple elements, and these elements are closely interrelated and form a kind of nexus security. For example, energy, food, climate, ecosystem, and environment all depend on water and have impacts on water. Such nexus nature of ecological security indicates mutual dependence and integrity of the elements.

Nature of market: The elements of traditional national security, such as arms, cannot be traded freely in the market. However, most of the elements of ecological security, for example, food, fossil fuels, and mineral resources, can be bought and sold freely in the market. Even water, an element whose supply is to a large extent restricted by spatial location, can be embedded in agricultural product trade, or directly traded as products, for example, in the form of mineral water.

Nature of gradually changes: In most cases, changes in the security of ecological elements do not happen abruptly. For instance, climate change is a long and gradual process; land desertification is not a phenomenon that happens in 1 or 2 days nor just in 1 or 2 years.

Nature of sovereignty territory: All elements of ecological security are linked to certain locations; hence, they clearly belong to certain sovereignty territory.

Nature of global integrity: Ecological security not only influences one region or one country, it has an impact on the mankind, and the whole world. For example, ozone holes, climate change, and the protection of endangered animal and plant species are all examples of ecological security issues that are beyond the border of any country.

Due to the speciality and diversity of ecological security, its security implications are very different from traditional security, which requests in-depth and accurate interpretation.

There are three major types of factors that influence ecological security. The first one is natural factors. As ecology is the relationship between human and the nature, human beings should always recognize, follow, and accommodate the law of nature. Temporal and spatial changes of precipitation and the seasonal, interannual, diurnal variations in temperature are beyond the control of human power. If the relationship between human and nature is in harmony, there won't be any ecological security risks. The problem is that the nature is in constant change and sometimes such changes are in the form of extreme events, for instance, droughts and floods. Meanwhile, human society needs stability and proper natural environment. Human society has to adapt to the changes and variations of nature. The supply of some natural resources, for example, fossil fuels, which takes millions of years to form and the reserves are limited, can't be indefinite.

The second type of factors is market factors. As some elements of ecological security are market goods, their market security is crucial to ecological security. For fossil fuels, apart from the limited reserves, in certain time period, even if there is market supply, it does not necessarily mean the satisfaction of energy demand. If the price of fossil fuels may be too high and beyond the consumers' affordability, then the demand cannot be met. In such situations, there is no food and oil security. In some markets, for example, food market, some speculators control the supply and artificially raise the price, making it unaffordable to ordinary people. Therefore, food security is not decided by food production but by the market power. Furthermore, even if the market price is reasonable, there are issues whether the grain import can be successfully transported home and reach final consumers. If the transport market does not function properly or is unsafe, the fossil fuels and food a country buys in the international market may not be able to reach the final consumers. In that sense, the market factors of ecological security is not only about whether there are goods supply in the market but also about whether these goods are affordable and can be transported to final consumers.

The third type of factors is human ones. Human factors include two kinds, out of malice or no malice. Take greenhouse gases emissions as an example, the emissions are not out of malice. Moreover, some of the emissions are for basic needs or substance. Natural system could absorb some greenhouse emissions and neutralize them. But when the emissions exceed the capacity of nature's self-purification capacity, these emissions could transform from unintended side effects to malicious emissions, for example, luxury or wasteful emissions. Subjective malice by human beings is manifested by targeted interventions of the human beings to the natural system, which leads to major security threats or huge losses. Examples of such behaviors include manipulating the global oil price out of subjective malice, monopolizing the global food market for profit maximization, and controlling important natural resource for strategic reasons. Besides, some of the huge projects crucial for the life of large numbers of people, for example, the South–North Water Diversion Project and the

Three Gorges Dams, can also be targets of malicious damage actions. Damages to these projects can affect ecological security in large areas.

It is important to recognize the importance of ecological security. China has a large population, a big economy, and a vulnerable ecosystem. It lacks mineral resources and is highly integrated into the international market. It highly depends on the international import for the supply of energy and some other natural resources that are crucial to the normal function and development of Chinese economy. In 2013, China's crude oil production was 209 million tonnes, while the import of crude oil was 282 million tonnes, and the import of refined oil was 39.59 million tonnes. This means over 60 % of China's oil consumption depends on import. For a long time, it has been believed and stated that China has abundant coal resources; however, in recent years, China's import of coal has been growing rapidly. In 2013, China's raw coal production was 3.68 billion tonnes, and its coal import was 330 million tonnes. In the same year, China's crude steel production was 780 million tonnes, and its import of iron ores and the concentrated iron ores was 820 million tonnes. Besides, China saw another year of good harvest in 2013, and its total grain production reached 600 million tonnes. However in the year, it still imported over 80 million tonnes of grain, including 63.4 million tonnes of soybeans, 14.6 million tonnes of cereal, and 8.1 million tonnes of cooking oil.[13] These are resource import; meanwhile, they are also imports of embodied water and embodied energy and have important ecological security implications. For years, China's iron and steel production make up 45 % of the global total, and its consumption of some metals is around 60 % of the global total (see Table 5.4) (Fig. 5.4).

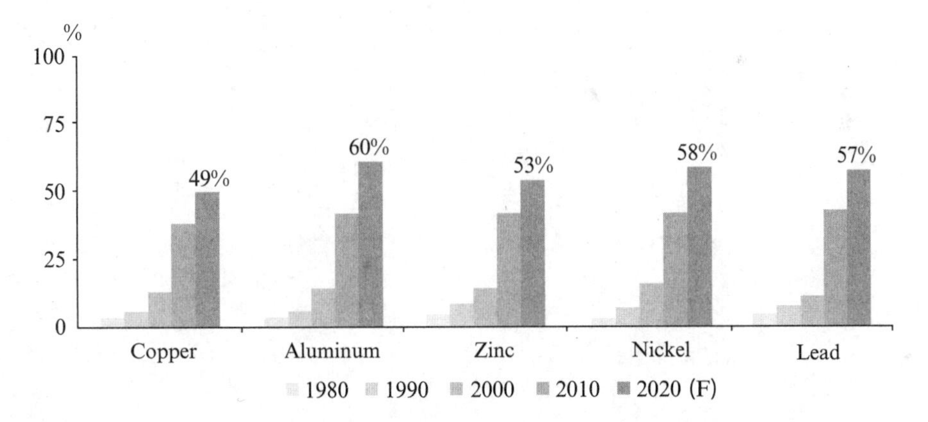

Fig. 5.4 Proportion of China's metal consumption in global total, 1980–2020 (Data taken from Lennon, Jim (2012), Base Metals Outlook: Drivers on the Supply and Demand Side, presentation, February 2012, Macquarie Commodities Research, www.macquarie.com/dafiles/Internet/mgl/msg/iConference/documents/18_JimLennon_Presentation.pdf)

[13] China National Statistical Bureau (2014). *2013 National Statistical Communique on economic and social development.*

Three Gorges Dams are also [illegible] of [illegible] damage actions. Damages to these projects can affect ecological security in large area.

It is important to recognize the importance of geological security. China has a large population, a [illegible] economy, and a [illegible] ecosystem [illegible] mineral resources and is firmly integrated into the international market. It highly depends on the international market for supply of energy and some critical mineral resources that are crucial to the economic growth and development of China's economy. In 2013, China's crude oil production was 209 million tonnes, while the import of crude oil was 282 million tonnes, and the import of [illegible] was [illegible] tonnes. This accounts for 60 % of China's [illegible] consumption [illegible]. For a long time it has been believed [illegible] that China has abundant coal resources. However, in recent years, China's import of coal has been growing. [illegible] 2013, China's [illegible] production was [illegible] billion tonnes, and [illegible] was 330 million tonnes. In the same year, China's [illegible] production was [illegible] and [illegible] was [illegible] million tonnes. Besides, China's [illegible] in 2013 [illegible] production [illegible]. However, in the [illegible] it still imported over 80 % [illegible] million tonnes of soybeans. [illegible] These consumption [illegible] water and [illegible] energy, which have [illegible] ecological security [illegible]. For years, China's [illegible] production has [illegible] up to 45 % [illegible] global total and its consumption of some metals is around 50 % of the total (Fig. 5.4).

Fig. 5.4 [illegible] consumption [illegible] global total [illegible] 2020 [illegible] China [illegible] Beijing: [illegible] University [illegible] Metal [illegible] Commodities Research [illegible]

[illegible] China National [illegible] Bureau [illegible]

Chapter 6
Low-Carbon Energy Transformation

The energy used under industrial civilization development paradigm is the fossil fuel. However, the limits of fossil fuel reserves show that if the ecological civilization paradigm pursues sustainable development, it has to find sustainable energy resources. In the view of the general patterns of world energy consumption, fossil energy provides the basic energy support in the process of industrialization and urbanization. However, the future energy demand far exceeds the exploitable reserves of fossil energy. Transformation to renewable energy is inevitable and also revolutionary. China's ecological civilization construction should focus on speeding up this transformation.

6.1 Energy Consumption Patterns

Before the industrial revolution, human society and economy were based on the agricultural civilization, and the energy consumption was mainly small scale and scattered renewable biomass use, and coal was mainly used locally and at small quantity to daily energy needs of households. Because of its low energy content, big volume, and collection difficulties, biomass could not meet the energy need of mass industrial production; therefore, the exploitation and utilization of coal continued to expand and the share of biomass in energy consumption declined. As the population kept growing after in industrial society, biomass energy was used as primary energy in the traditional agricultural society for a long time. Industrial production capacity and scale expansion were stopped and reversed during the two world wars. However, after World War II, with the acceleration of industrialization, coal consumption exceeded the traditional biomass use. Moreover, due to the rapid development of the automobile industry, the production and consumption of petroleum, as a more efficient and higher-quality source of energy, grew rapidly. In the late 1960s, petroleum replaced coal as the world's largest commercial energy. In the 1970s, the composition of world's energy production and consumption is that fossil energy,

J. Pan, *China's Environmental Governing and Ecological Civilization*,
China Insights, DOI 10.1007/978-3-662-47429-7_6

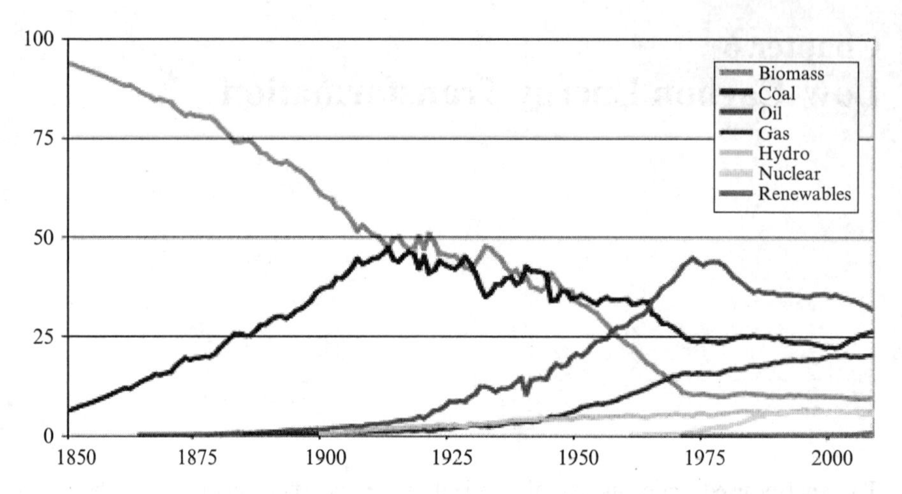

Fig. 6.1 Change of World Energy Consumption Composition from 1850 to 2010 (Source: World Energy Assessment Report, 2013)

including petroleum, coal, and natural gas, accounted for 80 % of total energy consumption, while the other 20 % came from biomass and hydroelectric and nuclear power (see Fig. 6.1). Meanwhile, apart from biomass, other renewable energy sources, such as wind and solar energy, were still in the initial development stage. The structure of global energy consumption in 2012 is coal accounted for 29.9 %, petroleum 33.1 %, gas 23.9 %, nuclear power 4.4 %, hydropower 6.7 %, and other renewable sources 2 %. Compared with the global energy consumption composition, China's energy structure is of lower quality and higher carbon content and needs revolution change to transform the coal-dominated energy structure to a low-carbon structure.

On the whole, China's energy consumption composition shows the same trend with the world but also has its own features. First, the large-scale use of fossil energy lags behind the developed countries, which is because of the late start of industrialization and urbanization in China. Second, a larger proportion China's energy consumption comes fossil energy. Fossil energy accounts for more than 90 % of China's energy consumption; moreover, coal dominates China's energy mix, accounting for about 68 % of the country's total primary energy consumption, petroleum less than 20 %, and natural gas is only less than 5 %. This is mainly due to China's energy resource endowment of small petroleum and natural gas reserves. The development of nuclear power started late in China, and the scale of nuclear power is small; in contrast, hydropower development began early and is of much larger scale. Hydro accounts for 8 % of China's energy mix, while the share of nuclear power is around 1 % and that of other renewable energy less than 2 %. Therefore, it can be seen that China's energy mix is of high-carbon content, high pollution, and low efficiency. Due to rapid urbanization and lifestyle change of rural households, the direct combustion of traditional biomass has reduced sharply. Traditional biomass only contributes a very small share of China's energy consumption, and this noncommercial energy form is often neglected in national energy statistics. Despite the rapid development of wind, solar energy, and other renewable energy use, their share in China's commercial energy mix is still tiny. After several decades of efforts, China's energy consumption mix has undergone some positive changes.[1] In 2013, the proportion of coal in China's primary energy consumption was 67.5 %, a record low in recent Chinese history; petroleum accounted for 17.8 %, also the lowest proportion since 1991; natural gas was 5.1 %, double the proportion 10 years earlier. The proportion of nonfossil energy (including nuclear energy) increased to 9.6 %, with a growth rate of more than 50 % in the previous 10 years.

In 1949 when the People's Republic of China was established, China's total energy consumption was less than 30 million tonnes of coal equivalent (tCe); 30 years later, at the beginning of the reform and opening up, the national primary energy consumption reached 600 million tCe.[2] Since 2009, China's primary energy consumption has surpassed the United States and become the biggest in the world. From the perspective of energy production, in 2013, China's energy production was 18.9 % of the world total; in this year, China's energy production increased by 2.3 %, far lower than the annual average growth rate of 7.4 % during the previous 10 years. In terms of energy consumption, in the same year, China consumed 3.75 billion tCe of energy, accounting for 22.4 % of global total. As for growth rate, China's energy consumption grew 4.7 % in 2013, lower than the average annual growth rate of 8.6 % during the previous 10 years.

China's energy production and consumption have undergone some significant changes. The first change is the transformation from traditional biomass to fossil energy. Before the reform and opening up, the energy consumption of rural households was mainly the traditional biomass energy; in 1998, traditional biomass still contributed 57 % of the energy consumption of rural households.[3] The situation had changed fundamentally in 2014. Straw was once the major daily fuel for rural households. Now, many farmers burn straw in the fields after the autumn harvest, causing serious air pollution and becoming a seasonal cause of smog. Among the commercial energy, more than 90 % of the increase in energy production and consumption is fossil energy. However, China's energy production and consumption have not completed or realized the transition from coal to petroleum and natural gas. Another significant change is the change from a net energy exporting country into a net energy importing country. On the whole, China was still a net energy exporter in 1991: in the year, China exported over 20 million tonnes of coal and its coal import was less than 2 million tonnes, and its oil export was 28 million tonnes, while its oil import was about 10 million tonnes. Since then, China's energy import has exceeded its energy export. By 2013, China's net oil import had reached 320 million tonnes, and its net coal import was more than 300 million tonnes. The third major change is

[1] British petroleum company, 2014. BP world energy statistical yearbook.

[2] Zhou Fengqi. and Wang Qingy (edited) (2002). *50 Years of China's energy* (p. 8). China Electric Power Press.

[3] Zhou Fengqi. and Wang Qingy (edited). (2002). *50 Years of China's energy* (p. 401). China Electric Power Press.

the significant improvement in energy efficiency. In 2000, China's energy consumption for producing a tonne of steel was almost 0.8 tonne of coal, and the efficiency of coal-fired electricity generation was almost 400 g of coal for each kilowatt hour of electricity. At present, the energy efficiency for steel production and electricity generation has both increased by over 20 % and is close to or even exceeds the international advanced level.

6.2 Energy Demand

China's energy demand has become the largest in the world and is still growing rapidly. On the one hand, China's development stage determines that China's energy demand will continue to grow for a long period; on the other hand, due to the energy resource endowments in China, energy demand growth, especially the excessive growth of fossil energy, needs to be curbed to safeguard the country's energy and environmental security. The Chinese government has made many efforts to control the total national energy consumption and to improve national energy supply security and environmental security. Objectively speaking, all these efforts are effective; but the energy consumption growth has not slowed down. The underlying reason is that, in the absence of development paradigm change, efforts to change production patterns and lifestyle cannot achieve a satisfactory effect.

With the growth of the Chinese economy, the national demands for energy and power grew rapidly. In rural areas, the destruction of forests in order to get firewood caused damages to local ecosystems. To solve this problem, in the 1980s, the Chinese government vigorously promoted the development of small hydropower projects for rural electrification. As a result, the shortage of power supply in some water-rich regions in South China has been partially alleviated. Meanwhile, the government also encourages and supports the application of biogas among rural households to meet the daily energy needs in rural areas. However, due to the annual and seasonal fluctuations of hydropower supply and the limited production of hydropower, small hydropower stations cannot meet the energy demand in small and medium-sized cities and rural areas. Because of seasonal fluctuations in biogas production, poor production conditions, and lack of equipment management techniques, rural household biogas digesters rarely become the main energy sources for households in relatively rich areas. Despite the positive effects of small hydropower and household biogas digesters, they cannot fundamentally meet the energy demand of rural areas. Thus, under the current technical and economic conditions, it is difficult to meet the energy demand in the process of industrialization and urbanization with renewable energy.

Because of its big population and large economy, China's energy consumption has been accounting for a large share of the world total and continuing increase since China entered the development stage of rapid industrialization and urbanization. Even in 1971, before its reform and opening up, China's energy consumption was about 8 % of the world total. In 2000, this share increased to 11 %, which was about half the energy consumption of the United States. According to forecasts by the International Energy Agency, China's energy consumption will remain more than 20 % of the world total from 2015 onward. However, China's per capita energy consumption is only one-third of the world average. China's per capita energy consumption level did not reach the world average level until the early 2010s. In 1971, China's per capita energy consumption was only less than 1/16 of that of the United States, 1/8 that of the United Kingdom, and 1/5 that of Japan. After over four decades of rapid growth, China's per capita energy consumption is about 70 % of the United Kingdom, 60 % of Japan, and 30 % of the United States. The development trajectories of per capita energy consumption in developed countries indicate that energy demand increase slows down, becomes stable, or slightly declines after a country enters the postindustrial stage. However, during the stage of rapid industrialization, energy consumption usually grows rapidly. In 1960, Japan's per capita energy consumption was only one-third of the United Kingdom, but 10 years later, Japan's per capita energy consumption rapidly caught up with that of the United Kingdom and surpassed it in 2000. After that, the two countries' per capita energy consumption levels are more or less the same and show similar development trend. American per capita energy consumption has always been high and more than double the levels of Britain and Japan, which are of similar development level; but after entering the 1980s, the United States' per capita energy consumption has steadily declined and the decrease are faster than those of Britain and Japan (see Fig. 6.2).

According to the development paradigm of industrial civilization, even if China does not reach the same high per capita energy consumption of the United States, but follows the example of Japan and Britain in per capita energy consumption, China's energy demand will increase further increase by one third. That means China's total energy consumption may exceed 5 billion tonnes of coal equivalents per year in the next 15 years. In fact, the IEA predicted that China's total energy

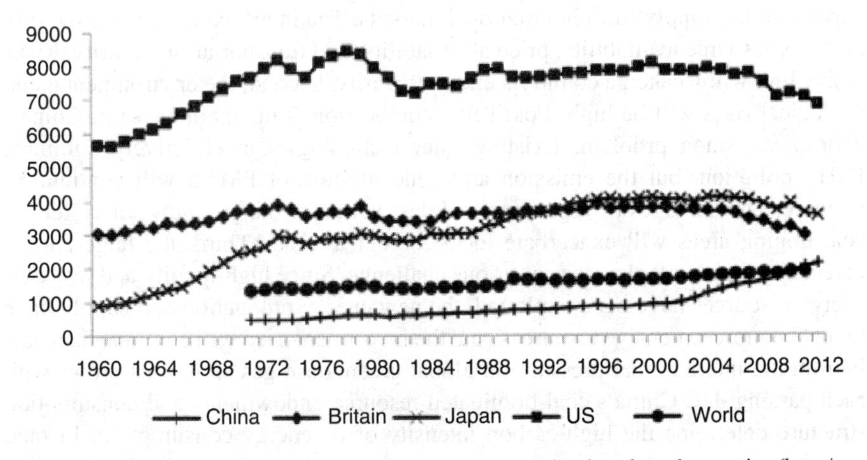

Fig. 6.2 Development trends of per capita energy consumption in selected countries (kgoe/person) (Source: World Bank database)

demand would reach approximately 5.5 billion tonnes of coal equivalent in 2030. By that time, China's urbanization rate will be between 68 % and 70 %, the services sector will account for 61.1 % of the national GDP, and industry's share of GDP will remain as high as 34.6 %.[4] In contrast, the urbanization rates in the United Kingdom and Japan have exceeded 80 % and 90 % respectively, and the shares of the service industry in their GDP are 79 % and 73 % respectively. Therefore, IEA's predictions about China's future energy consumption look reasonable. Currently in China, car ownership per thousand people is only 1/8 of the levels in the United Kingdom and Japan. Given that the energy consumption will increase only by 1/3 and the fuel efficiency will improve 50 %, this means that by 2030, China's private car ownership will only be about 1/4 of the levels of Britain and Japan. If China follows the development trends of Britain or Japan, even if China's per capita energy consumption could reach or exceed the levels of Britain or Japan by 2030, China's total energy consumption could further increase. If China will reach the same per capita energy consumption of the United States, China's total energy consumption could be more than 10 billion tonnes of coal equivalent after the completion of industrialization and entry into the postindustrial stage.

China cannot and should not follow the example of the United States in the energy consumption patterns. If China follows the energy consumption patterns of Britain or Japan, can its energy demand be satisfied? First of all, in view of China's energy resource endowments, the security of its energy supply cannot be guaranteed. In 2013, China's energy consumption was 3.75 billion tonnes of coal equivalent, the country's energy self-sufficiency rate was only 80 %, and oil self-sufficiency rate was only one third. If China's energy demand in 2030 will be one third higher than the current level, by 2030 China's annual energy consumption will be 1.25 billion tonnes of coal equivalent higher than present. China coal production has exceeded the level of safe production capacity; meanwhile, the possibility of further increase in China's proven oil reserves' growth is small. Although the production of renewable energy can rapidly increase, the additional energy demand will mainly depend on the supply from international markets. The interlinked resource security includes resource availability, price affordability, and transportation security. Risks in any link will weaken a country's energy security. Second, the environmental and ecological risks will be high. Fossil fuel combustion is the main cause of China's widespread smog problem. Existing filter technology can effectively eliminate PM10 pollution, but the emission and concentration of PM2.5 will continue to increase. The damages to vegetation and destruction of the groundwater system in coal mining areas will exacerbate local ecological risks. Third, the huge energy investment need will also pose a serious challenge. Since high-quality and low-cost energy resources have been exploited, the new energy production and supply have to rely on more expensive resources, and the investment need will continue increase. Fourth, the international pressure on global greenhouse gas emissions grows with each passing day. China's coal-dominated resource endowments and consumption structure determine the high-carbon intensity of its energy consumption. In fact,

[4] World Bank (2030) China.

China's CO_2 emission from fossil fuel combustion surpassed the United States and became the biggest among all countries in 2005. In 2013, BP statistical data on national annual CO_2 emission from fossil fuel combustion indicated that China's CO_2 emission accounted for 27 % of the world total, 60 % higher than that of the United States. China's per capita annual CO_2 emission is more than 7 tonnes, while the world average amount is less than 5 tonnes. As a responsible big country, China, although still a developing country, should contribute global greenhouse gas emission reduction.

Of course, developing nuclear energy can be an option. In fact, before the Fukushima nuclear station accident in Japan in 2011, China had an ambitious plan on the nuclear power development, which aimed at increasing the country's nuclear generation capacity from 10 GW in 2010 to 80 GW by 2020. Although the development of nuclear power has slowed down, China's installed capacity of nuclear energy is expected to reach 40 GW by 2020, which also indicates the severe challenges in the nuclear energy development. The fuel for nuclear power, uranium, is nonrenewable, and its reserve is limited. China needs to import large quantities of uranium ore to meet its demand. For the nuclear safety problem, investment and technology progress could enhance nuclear safety; however, nuclear power plants are operated by people; therefore, the risks of accidents due to human cannot be completely eliminated. As for the disposal of nuclear waste, currently nuclear waste can only be stored, and there is no way to treat nuclear waste safely. This is also the reason why Germany has decided to gradually give up nuclear power. As a transitional energy, nuclear energy is an important source of energy; however, its close connection to industrial civilization determines that it cannot be a sustainable energy source under the ecological civilization development paradigm.

European Union, the United States, Japan, and other developed countries followed the industrial civilization development paradigm and completed their industrialization and urbanization process on the basis of using fossil energy. The availability of fossil energy and the environmental and ecological risks China faces determine the difficulty of satisfying China's energy demand under the development paradigm of industrial civilization. China needs a new energy development paradigm and a new energy revolution.

6.3 Energy Revolution

Industrial revolution changed the path and course of human society development. A new technology formed a competitive advantage, generated huge market demand, and created enormous material wealth. Therefore, the industrial revolution was led by technology innovation; it was a revolution of spontaneous technology spreading and made the fruits of the technological revolution disseminate worldwide. So far, there have been many rounds of the industrial revolution, for example, many existing studies have identified electromagnetic technology, Internet technology, and three-dimensional printing as symbols of the second, third, and fourth industrial revolutions.

No matter how many rounds of industrial revolution have taken place, it does not change one thing: the power foundation of every industrial revolution is fossil fuel and other modern energy, not traditional biomass. It's hard to imagine that direct combustion of biomass can be the energy foundation for industrial revolution. Coal can be directly used to propel steam engines; however, to power TV and the Internet, it has to go through thermoelectric conversion, and only through gasification and changing into oil that coal can be used to fuel cars. It should be said that, nuclear technology, supercritical coal-fired power generation technology, wind power, and solar photovoltaic technology are all technological innovation, but they are not revolutionary technology breakthrough.

Under the background of increasing global concern about greenhouse effect and growing pressure on carbon dioxide emissions reduction, the concept of low-carbon revolution has received much attention. People hope that a low-carbon technology can create the revolutionary breakthrough of zero carbon emissions, thereby address the challenge of global climate change. People look forward to such a revolution, but they feel that such a revolution is still a distant dream. In international climate change negotiations, due to the lack of revolutionary technology breakthrough in low-carbon development, developed countries' emission reduction targets always much lower than expected, and most developing countries also emphasize their development priorities. Low-carbon revolution is in fact an energy revolution. What social economy development and people's daily life needs is energy, not carbon. If a zero-carbon energy revolution can take place, the ecological civilization development paradigm will have a sustainable energy base (Fig. 6.3).

We are possibly in the third or fourth industrial revolution now, and the new technologies include artificial intelligence, robot, digital manufacturing, and three-

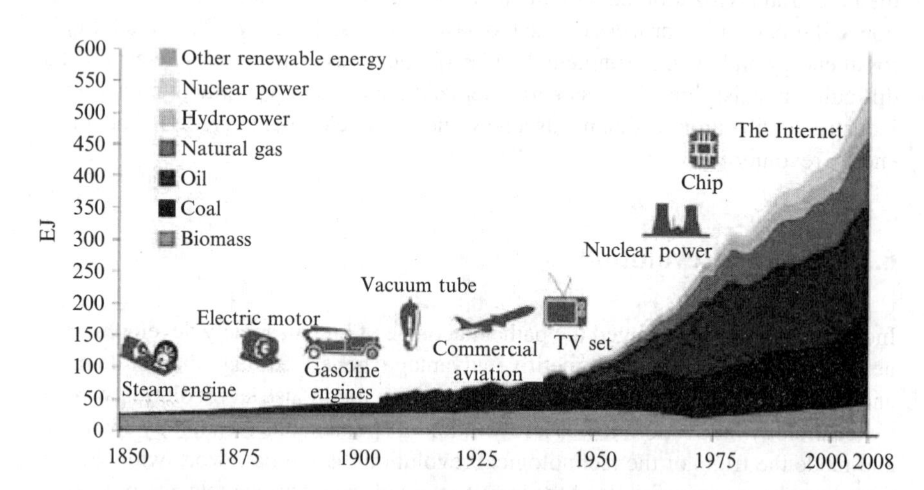

Fig. 6.3 Energy consumption and technological innovation

dimensional printing.[5] The current technology revolution depends on new energy sources, including shale gas and renewable energy. In fact, because of the breakthrough in shale gas technology, the United States' energy production pattern has fundamentally changed.[6] America's dependence on international energy supply has significantly reduced and may even disappear, and its energy mix has also dramatically changed, with substantial declines in the shares of coal and oil. However, there is no significant change in the production and consumption of renewable energy.

Does this mean that the shale gas revolution is an energy revolution? For the United States, shale gas is currently a technology breakthrough. But it should be noted that shale gas is also fossil energy and it is nonrenewable and exhaustible. Similar to other fossil energy, the mining and use of shale gas causes environmental pollution and damages to natural resources, which makes it less revolutionary. Shale gas production is not high enough to replace other main energy forms, and other energy sources are of an important or even dominant position. This means that even in the short term, it is difficult for shale gas technology to develop into an energy revolution in real sense.

More importantly, every technological innovation leading to energy efficiency improvement also requires a new form of energy. For example, the airplanes have to use kerosene and they cannot be fueled by coal. The Internet could only be driven by electric power, which could be generated from fossil energy or renewable resources. From the perspective primary energy use, the use of new technologies has not replaced or obsoleted traditional energy sources. For example, the production of iron and steel and cement still needs coal, oil, natural gas, and other fossil energy, in their primary form or converted form.

In this sense, there have been many manufacturing technology revolutions since the industrial revolution, but none of them is energy revolution in real sense. The sustainable development goals established by the United Nations could be expressed as "to ensure that all the people could have access to affordable, reliable and sustainable modern energy." The specific objective for renewable energy is to double the 2010 share of renewable energy in global energy mix by 2030 and increase it from 18 % in 2010 to 36 % by 2030. It also included the target of doubling the global speed of energy efficiency improvement, from the 1.3 % per year during 1990 to 2010 to 2.6 % per year by 2030. Technically, it calls for the strengthening of international cooperation to promote the research and development of clean energy technologies, including renewable energy, energy efficiency, and advanced clean fossil fuel technologies, meanwhile, promoting investment in energy infrastructure and clean energy technologies. It's thus obvious that the industrial revolution is a fact, while the energy revolution seems difficult to materialize. Despite the difficulties in global climate change mitigations, the international community does not dare to set any objective of "energy revolution."

From the perspective of world energy structure evolution and the exhaustible feature of fossil fuels, the energy form before the industrial revolution was renewable

[5] Washington Post (January 11, 2012) Why it's China's turn to worry about manufacturing.

[6] Huang Qunhui (2014).

energy, mainly biomass; after many countries started the industrialization process, biomass was replaced by fossil energy. Today fossil fuels dominate the energy mix of the world. Due to modern biomass energy technologies, traditional biomass is processed and supplied as a new form of commercial renewable energy; however, the share of modern commercial biomass energy is still very small. The industrialization process has lasted for 300 years, but only 20 % of the world's population have completed it. China is home to 20 % of the world's population, and its rapid industrialization on a large scale can potentially reshape the global landscape of energy consumption and greenhouse gas emissions, causing concerns about global energy security and climate security. If fossil energy could support global industrialization, human society will inevitably transit to renewable energy in the postindustrial era; if the limited reserve of fossil energy is not sufficient enough to support the current global industrialization process, the energy transformation will come early and happen during late stage of industrialization, or even earlier. It will not wait until the industrialization process finishes worldwide.

In fact, the energy use before the industrial revolution also included solar energy, hydro energy, and wind energy, though the main energy form was biomass. The big windmills in the Netherlands are an example of wind energy utilization. The drying of clothes and grain mainly relied on the light and heat of the sun. In ancient China, the mills driven by water flows were invented during the Jin Dynasty.[7] However, these are direct use of renewable energy, not modern renewable energy use. Obviously energy use under the ecological civilization paradigm cannot be simple and direct inefficient use of natural renewable energy resources. Whether it is forced transformation after fossil fuel reserve exhaustion or voluntary transformation before fossil energy depletion, the modern and efficient use of renewable energy is inevitable. The world needs an energy revolution, but the technology conditions for the revolution are not available and the revolution faces significant obstacles. Under such a background, how should we understand, recognize, and promote such a revolution?

Firstly, it is called a revolution because it is impossible to return to the traditional simple and extensive direct use of renewable energy, but it has to be modern energy form and service through modern technology conversion. The renewable energy has to be of high efficiency, easy to use, and at acceptable price. Take hydropower as an example, the energy comes from the potential energy of hydraulic difference; however, through electromagnetic transition, the turbines covert the potential energy into electricity.

Secondly, the revolution of renewable energy is hard to realize through its own energy production. It needs to be propelled by fossil energy. During the transformation stage, the combustion of fossil energy continues to provide for the manufacturing of water turbines, the production of reinforced concrete for the building of hydropower dams, and the production of monocrystalline silicon for PV power generation. Of course, after the completion of the renewable energy revolution, or after the depletion of fossil energy, the fossil fuel propelling will be no longer needed.

[7] http://b2museum.cdstm.cn/ancmach/machine/ja_21.html.

However, during the transformation from industrial civilization to ecological civilization, the technologies, energy base, and production patterns of industrial civilization could also play a positive role. In this process, the use of fossil energy and nuclear energy will facilitate the transformation.

Thirdly, renewable energy technology won't be a single technology. The technologies that industrial revolution relied on are mostly the spreading and deviation of a single technology after technical breakthrough. Renewable energy technology includes biomass, wind energy, solar energy, geothermal energy, and tidal energy, and these renewable energy resources could be utilized in multiple forms, such as heat and electricity. If limited to a single type of renewable energy, the amount of commercial energy acquired cannot meet the needs of social and economic development. Therefore, renewable energy revolution is the technological innovation regarding various renewable energy resources.

Fourthly, in addition to energy utilization technologies, energy revolution also includes other related technologies, such as energy storage technology and smart grid technology. The high heat value of fossil fuels determines their storability. However, renewable energy is different and it varies from time to time. Wind energy and solar energy cannot be stored and transported as fossil energy, and they need to rely on electricity storage technology. Biomass stores a lot of solar energy and could be stored directly, but its production is highly affected by the weather, and the heat content of each cubic meter of biomass is lower than coal and oil. Hydropower can be stored through the construction of dams; however, water availability is also subject to seasonal and annual changes. Smart grid can be precisely scheduled and controlled; hence, its design can take the form of "energy Internet."

Fifthly, renewable energy revolution is not only the revolution of commercial energy but also includes noncommercial energy. Solar thermal drying is a form of noncommercial energy. Solar energy water heater suppliers clearly do not directly sell energy, but equipment for energy conversion or harvesting. If the energy services could be obtained through direct harvesting or conversion of renewable energy instead of purchasing commercial energy on the market, then the noncommercial energy service is equivalent to the commercial energy services it replaces or realizes. Other examples include solar PV for community or street lighting, mobile phone or computer recharging, distributed home power systems, and small rural biogas digesters. To some extent, greenhouses are also equipment or method for direct harvesting of renewable energy. Many studies and analysis regard such noncommercial energy use as energy efficiency practices, which is actually wrong.

Sixthly, the renewable energy revolution is a revolution in energy consumption. It's generally believed that consumption is a social behavior, not a technology. Energy revolution should be understood from the technical perspective. However, if these technologies are not accepted by consumers and cannot stimulate conscious consumer behaviors, the technology revolution will not be successful. People can spend large sums of money to buy luxury villas and nonessential consumer goods such as diamonds and antiques, but they are reluctant to spend money to install the photovoltaic power generation panels on their villas. From market and asset perspective, such behaviors are of market rationality. Because of their scarcity, the values of diamonds and antiques will increase over time; while photovoltaic devices are not scare, and because of technological progress, the market price of these devices will decrease, and these devices need regular maintenance and update. However, if hundreds of millions of consumers use the solar energy water heaters, solar photovoltaic for lighting, mobile phone recharging, or electric car recharging, what will be the effect? Some say it is energy saving. In fact, these energy services are not energy saving, they are effective consumption. Consumers shift from commercial energy service consumption to the consumption of devices for non-commercial energy collection. In a crowded city, a luxury sports car, except for its luxurious appearance, has the same performance as ordinary cars. If people use an electric car, the behavior can stimulate similar behavior among other consumers and lead to a consumption revolution, causing revolutionary change in the quantity, quality, and variety of energy demand. Fashionable clothes are in fashion for a short time, and seldom their values increase over time. But they are very powerful in stimulating innovation. Even in absolutely insulated houses, people may choose to keep their air conditioners running while their windows are open; in such cases, the energy demand and consumption is because of consumer behaviors, and they cannot be addressed with technology innovation. The so-called rebound effect is consumers' natural reaction to improvements in technical efficiency, which significantly reduces the actual energy-saving effect of technology progress.

Seventhly, we cannot expect a revolutionary breakthrough on energy efficiency technology, because efficiency improvement in energy technologies is a gradual process, and there is also a constraint on the feasible space. For example, the potential for further improving the efficiency of thermoelectric conversion technology in ultra-supercritical conditions will be smaller and smaller. Moreover, energy conversion rate cannot exceed 100 %. Technology itself contains uncertainty and even causes some side effects. Take building insulation materials as an example. In the absence of relevant quality standards, some combustible material may be used as insulation materials and causes fire. Some technology progress is purely for profit purposes, for example, the technology progress of several new generations of mobile phones could be combined into one generation, and the generation change could be much slower. However, the mobile phone makers, to maximize their profits, constantly put new models, new settings, and new applications of mobile phones into the market, making the old and used mobile phones, which consume lots of energy in their manufacturing process, become outdated and wasted before the end of their use life.

Eighthly, institutional setup and administration system design are probably the most important influencing factor for energy revolution. All the seven items mentioned above need some institutional arrangements to facilitate and safeguard the corresponding conditions for the revolutionary technology breakthrough. If people continue following the doctrine of utilitarianism and targeting profit maximization and material wealth accumulation, it will be impossible to achieve the necessary revolutionary breakthrough in energy technology. On the contrary, if people respect the nature, pursue the harmony between human and nature, and use sustainability as development measurement, the technology orientation, price system, and consumer preferences will facilitate the realization of energy revolution.

6.4 The Transformation Practice

The energy revolution has not succeeded, and various efforts from all aspects are still necessary. Industrialization, urbanization, environmental pollution, and resource shortages are directly related to the combustion of fossil energy, and fossil energy cannot be replaced on a large scale in the short term. Under such circumstances, various policies, technologies, and even slogans are urgently needed to promote the energy revolution practices, such as stimulating the development and application of resource- saving, environment-friendly, energy-saving, green, recycling, livable, low-carbon, and smart technologies.

The effectiveness of all these practices could be measured with three parameters, which are quality, energy efficiency level, and cost. Forged and fake commodities and jerry-built projects are obviously "resource saving," low cost, and high return. But because of their poor quality and short use life, energy use and cost for their production are often wasted, and the long-term losses are bigger than the immediate gains. In many house construction projects, in order to save cost, stronger concrete reinforcing bars in the original design are replaced by weaker and cheaper ones, and low-grade cement is used despite the original design that requires the use of high-grade cement. Because the construction quality does not reach the relevant standard, the buildings either need to be reinforced or demolished for reconstruction. Such practices lead to waste resources, investments, and energy resources. Resource conservation depends on cost and energy consumption. Resource conservation could be achieved at the price of high costs and energy consumption. Moreover, poor construction quality may generate high profits for the building companies because the construction, demolishment, and reconstruction all cost money and generate profits, and the companies involved may be able to recover their investment. However, the fossil energy consumed cannot be recovered. In this sense, the ultimate indicator of successful transformation is energy saving.

Similarly, there is no obstacle for recycling, green, and smart technologies and practices; there are no barriers to them under the development paradigm and technological conditions of industrial civilization. In space stations, water is recycled completely. Existing technologies can treat and purify industrial wastewater and household wastewater and make the purified water satisfy drinking water quality standard. In the Gobi Desert, the sea water could be desalinated to irrigate trees and grasses. An automated intelligent system can be established as long as there is investment. All of these constraints seem to be capital investment at first look, but the actual constraint is energy. The natural supply of sustainable energy is not abundant enough to support the realization of artificial recycling and desalination of sea water. In the absence of energy conservation, smart technologies are not really smart. Thus, the ultimate criterion of green, recycling, and smart technologies and practices is still energy efficiency.

During the 2006–2010 period, energy conservation and reduction of pollutants were included as mandatory requirement for the 11th Five-Year Plan, requiring that energy intensity (energy consumption per million yuan of GDP) in 2010 should decrease by about 20 % and the emissions of chemical oxygen demand (COD) pollutant and sulfur dioxide should be reduced by 10 % compared to the 2005 levels. The Energy conservation target is a relative target, the improvement of energy efficiency is mandatory, while the total energy consumption can continue to increase. The pollutant reduction targets are absolute targets, requiring that the absolute amount of pollutant emissions must be reduced regardless of emission intensity change. During the 2006–2010 period, the central government ordered the closure and eliminating of inefficient and polluting production capacity for achieving the targets. As a result, the energy consumption per million yuan of GDP declined by 19.06 % during the 2006–2010 period and the national emissions of COD and sulfur dioxide reduced by 12.45 % and 14.29 % respectively. That means the environmental pollution reduction was more effective than planned, while the energy efficiency improvement was lower than the target of 20 %. The reason is actually quite simple. With investment and energy consumption, the pollutant emission can be reduced sharply. Take desulfurization as an example; as long as enterprises invest in and run desulphurization equipment, sulfur dioxide emissions can be effectively controlled. However, the operation of desulphurization equipment needs to consume electricity. Therefore, the National Development and Reform Commission offered some subsidies to enterprises to help them cover the cost of the additional energy consumption.

Sewage treatment is of similar situation. The construction of the sewage treatment plants need investment, while keeping them running needs energy—sewage treatment process consumes electricity.

The key issue and focus of energy transformation is renewable energy use. If renewable energy could provide more energy services than the total economic and social demand of a country, energy saving won't be necessary. A simple example is household solar energy water heaters, which often provide more than enough hot water in summer for the ordinary family but not during winter, leading to waste of heat energy. China's energy transformation started early and is in harmony with nature. Due to China's late start of industrialization and lack of capital and technology, China's energy production cannot meet the need for social and economic development and causes lots of damages to the environment. Back in 1958, Mao Zedong encouraged the development of rural biogas digesters and pointed out that "biogas can be used for lighting, cooking, and generating fertilizer; biogas development should be vigorously promoted."[8]

By 2005, China had built 18 million biogas digesters. From 2006 to 2010, the Chinese government offered subsidies to the building of 13.2 million household biogas digesters; the total investment was 40.1 billion yuan, among which the fiscal subsidies from the central government was 12.5 billion yuan.

[8] *Construction Plan on National Rural Biogas Digesters from 2005 to 2020.*

According to estimates by the Chinese Ministry of Agriculture, there were about 40 million household biogas digesters in China in 2010, providing 15.4 billion m^3 of biogas, which is equivalent to 24.2 million tonnes of coal. In 1983, China carried out the pilot project of small hydropower construction at the county level and promoted the development of medium and large hydropower projects,[9] but the commercial energy supply was not enough to meet China's energy needs for national economic and social development.

Ecological civilization construction has been included in the Chinese government agenda in the twenty-first century, and China's renewable energy development has speeded up. In 1998, China had few modern commercial renewable energy productions which require advanced technology and large investment, such as the wind energy and solar photovoltaic power. At that time, the total installed capacity of photovoltaic power was only 13 MW, and the Chinese government set the target of increasing it to 300 MW by 2010. The actual installed capacity reached 800 MW in 2010, far exceeding the original target (see Table 6.1); the government further set the target of increasing the installed capacity to 21 GW by 2015. According to the *Renewables 2014 – Global Status Report*, China's installed capacity of photovoltaic power in 2013 has reached 19.9 GW. For the utilization of solar thermal energy, the collector area in 1998 was only 15 million m^2, increased to 80 million m^2 in 2005, and further 168 million m^2 in 2010, 12 % higher than the government target.

There are also successful practices in energy transition in developed countries. In the past three decades, Germany's total energy consumption has reduced by 15 %, meanwhile, its GDP increased by 80 %, realizing the decoupling of economic

Table 6.1 Transition toward renewable energy: practices in China

	1998	2005 (Actual)	2010 (Planned)	2010 (Actual)	2015 (Planned)
Hydropower (GW)		117.39	190.00	216.06	260.00
Wind (GW)	0.242	1.26	10.00	31.00	100.00
Solar PV (GW)	0.013	0.07	0.3	0.8	21.00
Biomass power (GW)	0.80	2.00	5.50	5.50	13.00
Biogas (m^3)	2.36	8.0	19.0	14.0	22.0
Household biogas digester (million)		18.00	40.00	40.00	50.00
Collector area of solar water heaters (million m^2)	15.00	80.00	150.00	168.00	400.00
Bioethanol (million tonnes)		1.02	2.00	1.80	4.00
Biodiesel (million tonnes)		0.05	0.20	0.50	1.00
Total (million tonne coal equivalent)	3.05	166.00		286.00	478.00

Note: 1998 data comes from Zhou Fengqi and Wang Qingyi, pp. 425–426, other data are from the 11th and 12th Five-Year Plan on Renewable Energy

[9] Zhou Fengqi and Wang Qingyi (2002) p. 424.

growth and energy consumption. Germany plans to stop nuclear power generation by 2022, continue to reduce the proportion of coal, and promote the production and consumption of renewable energy. According to its plan, by 2022, Germany's total energy consumption will be 20 % less than in 2008. By 2050, its total energy consumption will further decrease by 50 % from the 2008 level. Renewable energy will account for 18 % of the total energy end-use consumption in 2020, 30 % in 2030, and 60 % in 2050. The proportion of renewable power generation in the total electricity consumption will be 35 % in 2030 and 80 % in 2050. In general, energy conservation in transportation will be more difficult. However, Germany has also made an ambitious plan of reducing its total energy consumption in transportation in 2020 by 10 % from the 2008 level and 40 % in 2050. Today in Germany, solar PV panels are installed on the roofs of many residential buildings and houses, and the electricity generated can directly access the power grid. The United Kingdom, the birthplace of the industrial revolution, has released plans on speeding up, reducing its fossil energy consumption. The United States also announced in 2014 to reduce the carbon emissions from its power sector by 30 % by 2020.

The practices on energy transition in different countries indicate that in none of countries, the transitions happened spontaneously due to a leading technology like the industrial revolution; instead, they have been realized through the arrangement of institutional mechanisms. The policies and measures taken to promote and guarantee energy transition include legislation, standards, subsidies, and so on. In Germany, for example, government subsidies cover 50 % of the total cost of having solar photovoltaic panels installed on residential buildings and houses; in addition, the electricity surplus can be sold to the grid, and the tariff is higher than the price of electricity charged to consumers. On the one hand, the practice of energy transition is passive and mandatory, unlike the spontaneous spreading of steam engines, automobiles, the Internet, and other technologies during the industrial revolution which did not need government subsidies and international agreement. On the other hand, such unspontaneous change is an active pursuit, because its clear direction and objectives and effective measures will promote the transformation of civilization paradigm.

6.5 Transformation Strategy

The energy demand of the process of urbanization and industrialization is lasting and enormous. The limited reserves and pollution of fossil energy force people to accelerate energy transformation. The potential safety problems of nuclear power make countries hesitant in nuclear power development. The low heat value, intermittence, and high costs of renewable energy make people doubt about their role and lack confidence in their development. Energy conservation is an effective solution, but for China, a big economy in the stage of rapid growth, the speed of energy efficiency improvement cannot be as fast as the increase of energy demand. Theoretically speaking, the growth in energy demand will continue until China becomes a developed economy and its population reaches its peak. In other words, the absolute quantity of China's energy consumption will continue growing until 2030 or later. To break the bottleneck energy development, China must break the traditional conception about narrow energy strategy and look for new ideas and broad strategy.

The first solution is to diversify the energy structure. The focus and center of China's energy strategy has been commercial energy, especially fossil energy. Debates about energy security strategy mostly concentrate on oil security. However, in face of the international background of global climate change and the domestic need for ecological civilization construction, the traditional conception about energy strategy is unable to meet the actual needs of China's energy development. The 2015 International Climate Agreement requests countries to announce their Intended Nationally Determined Contributions and mitigation targets; domestic pollution control and ecosystem restoration have become rigid constraints on energy development; economic restructuring and improvement in the people's living standard demand both quality improvement and quantity increase for energy services. All these factors make energy mix diversification an urgent task. Commercial and noncommercial energy sources should be both valued, and the variety and quantity of renewable energy should be improved comprehensively. What the society needs is energy service, regardless of whether the energy service is obtained through purchase of commercial energy or direct use of energy from nature. If hundreds of millions of families and tens of thousands of schools and hospitals start to use solar energy water heaters, solar photovoltaic lighting, charcoal from biomass, and biogas digesters, the resulting effects will be far cleaner, safer, and more sustainable than several nuclear power plants and fossil fuel-fired power plants.

Second, energy administration should be integrated and coordinated. As a macroeconomic issue, the energy administration authority in China is the National Development and Reform Commission and the Energy Bureau. But in practice, energy administration is scattered among multiple government agencies and even some large state-owned enterprises. For example, small hydropower is administrated by the government agencies on water resources; biomass is administrated by the government agencies on agriculture; electricity, petroleum, and coal administration are mainly administrated and supplied by a few huge state-owned enterprises. Solar energy water heaters are administrated by government agencies on industry and information technology; and due to their market competitiveness, industrial or local government agencies rarely provide any subsidies or other support. The development of wind energy involves wind turbine manufacturing, site construction, and access to power grid and needs the administration and coordination of more government agencies. In order to formulate and implement an effective energy strategy, energy administration must be comprehensively consolidated, integrated, and coordinated.

Third, energy sector should be liberalized. Energy sector is an economic sector that not only provides products and services but also provides employment and ecological services. The production of fossil energy is technology and capital intensive, and the need for labor force is small. But the production, installation, and maintenance of renewable energy equipment demand a large workforce. The utilization of renewable energy has small impact on the natural ecosystems, and in some cases, renewable energy utilization could also provide ecological services. Therefore, China's energy strategy needs to include energy sector liberalization.

Fourth, the energy technologies should be diversified. China's energy technology strategy emphasizes on fossil energy, nuclear energy, and hydropower, and there are corresponding research institutes on their utilization. However, investments in the research and development for other nonfossil energy technologies have been small for a long time. In other words, government investments in the research and development of energy technologies concentrate in commercial energy and ignore noncommercial energy. If China gives the same research and development support to noncommercial energy as to commercial energy, even if the government investment is only a small share of that for commercial energy, major technology breakthroughs may be made in the scale and efficiency of solar energy water heaters and biogas digesters. Government support for the research and development of energy technology must be diversified, covering commercial energy, non-commercial energy, fossil energy, renewable energy, energy equipment technology, and energy efficiency technology. In summary, the development of energy technology must be comprehensive, diversified, and have multiple focuses.

Fifth, the energy market must be globalized. Energy technologies, products, and services are important content of the global economic integration. Because of its small fossil energy reserves, China needs to utilize both national and international resources and markets to guarantee the energy supply for its urbanization and industrialization. Even without the international requirement for greenhouse gas emission reduction, China has to reduce its fossil energy combustion to reduce environmental pollution and protect its ecosystems. On the other hand, the functioning of China's large economy and the safeguard of people's livelihood require a large quantity of products with high-energy embodiment. This means that the globalization of China's energy market not only involves energy products, technologies, and services but also includes products with high-energy embodiment, such as steel and electrolytic aluminum. The import of such products with high-energy embodiment can reduce domestic energy demand and emissions of pollutants. That is to say, China's energy strategy also needs to be open and cover industrial processes and product life cycles.

Energy development is both a challenge and an opportunity. Turning challenges into opportunities needs to change the strategies of energy development, improve energy and climate security, and promote the construction of ecological civilization.

Chapter 7
Sustainable Steady-State Economy

Since China's reform and opening up, the continued rapid growth of its economy has led to significant increase in both the size of Chinese economy and per capita income. The rapid growth attracts lots of interests in the future of Chinese economy at home and abroad. There are many different opinions, and no unanimous conclusion can be drawn. Some people are optimistic and think Chinese economy will continue its rapid growth; there are also pessimists who believe it will soon collapse, while others think it will go through some shifting and adjustment. Under the development paradigm of ecological civilization, China's economic growth is impossible and unnecessary to follow the growth path of industrial civilization development paradigm, and the growth transformation is inevitable. Accommodating nature means respecting the boundary constraints for the harmony between human and nature and avoiding any effort to maintain growth or promote growth at the cost of violating the laws of nature. Moreover, economic growth under the ecological civilization development paradigm must be real and in line with ecological harmony. Therefore, the future of China's economic transformation has to be structure adjustment, quality upgrading, and transition toward the steady-state economy of harmony between human and nature.

7.1 The Trend and Drivers of Economic Growth

China's economy had experienced some big ups and downs from the founding of the People's Republic of China in 1949 to the early 1970s, from shrinking by 27.3 % in 1962 to an annual growth of 18.3 % in 1965 and 19.4 % in 1971. It had some small fluctuation and maintained medium speed growth during the 1970s. From 1980 to 2010, the Chinese economy experienced a sustained explosive growth with an annual growth rate above 10 %. What the future of China's economy will be?

In 2005, Justin Yifu Lin, chief economist of the World Bank, stated in an article that the Chinese economy could maintain an annual growth rate of 8–10 % for

J. Pan, *China's Environmental Governing and Ecological Civilization*,
China Insights, DOI 10.1007/978-3-662-47429-7_7

another 30 years, and it would surpass the American economy in 2030 to become the world's largest economy.[1] In 2010, under the proposal of the World Bank, the research project "China in 2030" was jointly launched by the World Bank and the Chinese government. The research under this project concluded that the average annual growth rate of Chinese economy will be 6.6 % in the next 20 years, a third lower than the average annual growth rate of 9.9 % during the previous three decades. By 2030, such growth will make China become a developed country by 2030 and its size bigger than the American economy.[2] According to the baseline scenario established by the State Council Development Research Center of China, Chinese economy's average annual economic growth rate will be 6.6 % from 2010 to 2020, 5.4 % from 2020 to 2030, 4.5 % from 2030 to 2040, and 3.4 % from 2040 to 2050.[3] The 2050 Chinese economic growth scenarios set by QuanBai et al. (2009) show that China's average annual economic growth rate will be as high as 8.0 % from 2010 to 2020, 6.0 % from 2020 to 2035, and 3.8 % from 2036 to 2050.[4] Generally, foreign research institutions are slightly more conservative in their forecasts about the future of China's economic growth. For example, the International Energy Agency predicted that the average annual growth rate of China's economy would be 5.7 % from 2000 to 2010, 4.7 % from 2010 to 2020, and 3.9 % from 2020 to 2030. A study by the Goldman Sachs in 2003 concluded that the average annual growth rate of Chinese economy would be 4.35 % from 2015 to 2030 and 3.55 % from 2030 to 2050.[5] A representative of the pessimistic about China's future is Gordon Chang, who claimed that the Chinese economy was on the verge of collapse in his book in 2001.[6]

According to the World Bank, China's economic volume was the 12th biggest in the world in 1980, smaller than that of India; in 1990, its rank became the 11th, barely surpassing India; in 2000, China's rank jumped to the sixth, and then in 2010, it surpassed Japan to become the second largest economy in the world. China's per capita GDP was around the 150th in the world in 1980; by 2000, it was still the 136th. By 2010, it was almost the 100th, and 3 years later, in 2013, China's per capita GDP rose to the 75th in the world. However, compared to other countries, China's per capita GDP was still low. It was only 46 % of the world average, 12.16 % of that of the United States, and 14.22 % of that of Japan.

[1] Justin Yifu Lin. (2005). "China will surpass U.S. in 2030" *Southern Weekly*, Feb. 1 st.

[2] Work Bank-State Council Development Research Center Joint Research Team. (2011). *China 2030: Building a modern, harmonious, and creative high-income society.*

[3] MengkuiWang. (2006). *Important issues during China's mid-and-long term development*. China Development Press.

[4] QuanBai., YuezhongZhu., HuawenXiong., & ZhiyuTian. (2009). The future economic and social development of 2050 China. In *Energy and carbon emission report of 2050 China* (p. 893). Beijing: Science Press.

[5] QuanBai., YuezhongZhu., HuawenXiong., & ZhiyuTian. (2009). The future economic and social development of 2050 China. *Energy and carbon emission report of 2050 China* (p. 644). Beijing: Science Press.

[6] Gordan Chang. (2001). *The coming collapse of China*. New York: Random House.

In terms of GDP based on market exchange rate, in 2013, the US economy accounted for 22.43 % of the world total, while Chinese economy only accounted for 12.34 %, 10 % lower than that of the United States. If calculated on the purchasing power parity, the US GDP was 17.06 % of the world total, while China contributed 16.08 % of the world GDP, and the difference was only 1 percentage point. The British "Economist Intelligence Unit" predicted that the Chinese economy would surpass the US economy and become the world's largest economy in 2017 on purchasing power parity (PPP) basis.[7] In October 2014, the International Monetary Fund released its 2014 World Economic Outlook, and the data in this report showed that on PPP basis, Chinese economy was 200 billion dollars bigger than that of the United States in 2014, and China became the world's biggest economy 3 years earlier than the prediction by the Economist Intelligence Unit. However, based on market exchange rate, it is expected that China's GDP will catch up with that of the United States in around 2030.[8] Even in 2030, China's per capital GDP, even calculated on PPP basis, will only be 32.8 % that of the United States.

Except those who believed in the upcoming "collapse of China," studies at home and abroad all show that China has reached a transition period during which the economic growth speed will slow down, while the level of per capita income and national GDP will continue to increase. Direct or apparent reason for this transformation is change in the driving force for economic growth. Generally, it is agreed that China's economic growth has been jointly driven by three forces, including export, investment, and domestic consumption. China's opening up policy started from the coastal areas because both raw materials and the market were abroad and the coastal areas have the advantages of being closer to international markets. China jointed WTO in 2001 and was soon integrated in the process of global economic integration. With its strong competitiveness of cheap and high-quality labor force and favorable policies on land supply, foreign trade became the powerhouse of China's economic growth. To some extent, it is the export-oriented economy that stimulated the investment. Direct foreign investment led to an expansion of industries, and the massive investment in infrastructures made foreign trade more convenient. In the 1980s, there was almost no highway in China, the urban infrastructure was also insufficient, and the construction of sewage treatment system and underground rail transport in many cities had not even started. China's high saving rate and strong administrative power guarantee sufficient funding and high operational efficiency of energy, transport, and urban infrastructure constructions.

The domestic consumption's driving to the national economy has been relatively weak, which is closely related to China's special systems of urban–rural dualism, income distribution, and lack of a comprehensive social security system. As the traditional ways of governance are unable to provide the general public adequate

[7] The Economist Intelligence Unit, Morrison, W. M. (2013). *China's economic rise: History, trends, challenges, and implications for the United States*. US Congressional report, 21 Aug 2014.

[8] Arvind, S. (2013). *Preserving the open global economic system: A strategic blueprint for China and the United States*. Briefing PB13-16 of the Peterson Institute for International Economics. See *Pudong-American Economic Correspondence, 13*(347), 15 July 2013.

social security, people have to suppress their consumption urge, and traditionally, "thrifty" is regarded as a virtue. Thus, the economy growth has to rely on export. China's income distribution system shows multiple dual characteristics. In China, there exist multiple dualisms in income distribution. Among them, the most important one is the urban–rural dualism. The per capita disposable income of urban residents is more than three times higher than that of rural residents,[9] and in the calculation of this number, the hundreds of millions of migrant workers who are registered as rural residents but work and live mainly in cities are considered as urban residents. As less than 40 % of the Chinese population are registered as urban residents, the income level and consumption capacity of the rural residents, who make up over 60 % of the Chinese population, are very low. The second important dualism is between state-owned economy and private economy. State-owned enterprises enjoy market monopoly, and their formal employees enjoy high income, free health care, and cheap housing. In contrast, many private enterprises do not provide their employees with free health care and cheap housing security, and their employees' income level is less than one third that of the state-owned enterprise employees. As a result, employees of state-owned enterprises have low consumption demand, while employees of private enterprises have low consumption capacity. China is changing from a traditional agricultural society to a modern industrial society; however, the social security level for health care, education, pensions, and unemployment is low, and it only covers a small share of the population. Therefore, many people, despite their low income, try to save money for their own security.

American scholar Harry Dante believes that there is a correlation between population fluctuations and economic cycles. In this correlation, population change is the underlying cause of the change of the economic growth pattern, and the former determines the latter.[10] From the perspective of a person's life cycle, Dante found that the stage of a person's highest spending power is around 46 years old; therefore, a country's high consumption period is always 46 years later than its baby boom. From 1897 to 1924, the birth rate was high in the United States; 46 years later, the US economy experienced a period of rapid development from 1942 to 1968. The American baby boom started in 1937 and reached its climax in 1961, and from 1983 to 2007, the US economy enjoyed some unprecedented prosperity. The gap between the baby boom and the period of economic prosperity was also 46 years. After World War II, the Japanese economy underwent a period of rapid growth, which was known as the "Japanese miracle." It seemed that the Japanese government's industrial supportive policy helped protect Japanese enterprises from fierce international competition and expanded rapidly, create their competitive advantage in the export of high value-added commodities. But in fact, Japan's economic growth and rapid population growth are closely related. Because of its rigorous control on immigrants, Japan's economic boom was 46 years later than its baby boom. This

[9] According to the *2014 China Statistical Yearbook*, in 2013, urban residents' per capita consumption expenditure was 18,022.6 yuan, while rural residents' per capita consumption expenditure was 6112.9 yuan.

[10] Harry, D. (2014). *The demographic cliff* (trans: XiaoXiao). CITIC Press. Beijing, July 2014.

seems no coincidence. Demography indicates that human economic behaviors are cyclical. The data of the United States' long-term monitoring on 600 kinds of commodities show that young people get married at the average age of 26, which is associated with peaks in apartment rentals; these couples buy their first real estate property at the age of 31 for raising children; they buy their largest real estate property from 37 to 41 years old when their children are teenagers; their children's college tuition spending is at its peak when these couples are around 51 years old; after the children grow up and leave home, these couples will buy a luxury car when they are 53 years old. Forty-one years old is the time of debt peak, 42 years old is the year when their chip consumption reaches the highest level, and 46 years old is the year of the highest consumption during their lifetime. This means that the year of baby boom plus 46 years is the year when the next peak of economic prosperity will occur. Of course, this cycle is not absolute because massive immigration can also bring about rapid population change; technology progress can break this cycle, but it takes some time before it leads to changes in productivity; therefore, the impact of technology progress on economic growth is not immediate; increase in life expectancy can also postpone the consumption peak. Why does population growth lead to economic growth? Because population growth causes consumption growth, stimulate production expansion, promote market division and professional collaboration, and provide more opportunities for the applications of new technologies.

Obviously, the international market demand is not infinite, and it can vary and contain uncertainties as there exists an international competition. Investment is a long-term process, but the investment in infrastructure and housing, cars, and other durable consumer goods can reach saturation, and once it is saturated, the investment will become maintenance and refurbishment. Consumption will increase with the improvement of income distribution and social security, but it will also reach the ceiling as population peaks. The economic growth in the developed economies experienced a process of rapid growth, low growth, and then to steady state. China should take the initiative to make adjustments and adapt to the norms of the new economic growth stage, in order to avoid economic crisis and the "hard landing" of economic growth.

7.2 The Triple Constraints on Extensive Growth

The change of driving mechanism for China's economic growth transformation is not just because of the driving forces themselves, more importantly or essentially, it is because of the influences of external constraints on the driving forces. The external constraints of the harmonious development between human and nature are mainly in three aspects: natural factor, population factor, and capital stock.

To realize the harmony between human and nature, people have to first understand nature and then respect and accommodate it. Natural factors are a physical boundary, so technology progress can relax some of the tight constraints, but some other boundaries, such as the surface area of the Earth, are beyond the changing capacity of existing technologies. It is hard to imagine that people can change the structure and volume of the Earth. Even the most advanced and effective technologies, within a given range of time and space, the extent and rate they can relax the natural boundaries are also limited. Therefore, economic growth is subject to the rigid constraint of nature, which is the external pressure for economic transformation. Such pressure first of all comes from the deterioration and destruction of the living environment. In the 1950s, photochemical smog appeared in London because of air pollution, which was unimaginable in China in those days. Since the reform and opening up, many Chinese local governments believe in "no industries, no money" and support the development of polluting industries for rapid growth of local economy. Many parts of China did not set any restrictions on pollution when they attracted foreign investment and became "haven" for many foreign polluting industries. If the nearby water source is polluted, people just get water from a source further away; when the shallow groundwater runs out, people just drill deeper for phreatic water; when there is no deep phreatic water, people turn to long-distance water diversion; if the tap water is undrinkable, then people drink mineral water. Due to the limited production of mineral water, people use energy to purify the polluted water and produce bottled purified water and sell them as drinking water. Although drinking water becomes more expensive, people's income has also increased and can afford it, and they do not pay attention to the cost increase and water pollution. However, since 2011, serious smog have been appearing in the China's vast densely populated area, which forces people to change their attitude toward the rigid natural constraint of environmental pollution: air, which is more essential than water and its continual supply indispensable to human survival, cannot be bottled.

The second aspect of the natural constraint is the rapid price increase and reserve depletion of nonrenewable resources, especially fossil energy. Before the first oil crisis in the early 1970s, people were not aware of the problem of resource exhaustion. The soaring oil prices rang the alarm. However, people still think that the supply of nonrenewable resources is a price issue, which could be tackled through technology progress, discovery and exploitation of new reserves, efficiency improvement, and demand reduction. In fact, this has been the world response strategies during the past 40 years. As Chinese automotive industry was still in its early development stages, China was not alarmed to the soaring oil prices; instead, it considered large oil export as an opportunity for earning foreign hard currencies. Therefore, China, a country lack of oil resources, kept being a net exporter of crude oil until 1992. China used to think it has abundant coal reserves and exported large quantities of coal for a long time. Since 2010, China began to import large amount of coal. In 2013, China imported more than 300 million tonnes of coal. The exploitation of exhaustible resources is not just a problem of resource depletion; the collapse of coal mines and the destruction of groundwater systems have become a major cause of ecological disasters. Shale gas seems to bring about new hope to the development of fossil fuels, but the exploitation of shale gas needs to consume large amount of scarce water resources and causes severe groundwater pollution. Therefore, from the perspective of environment and resource, shale gas exploitation causes more damages than benefits. Continual technology progress, on the one hand, leads to constant improvement in resource utilization efficiency; however, because of the rebound effect and continued demand increase, the total amount of energy consumption continues to grow. On the other hand, advances in geological survey and mining technologies bring about the rapid approach of resource limit. The resource exploration tentacles have reached all corners of the land and ocean, and some low-grade mineral has also been exploited.

The third aspect of the natural constraint is the limited availability of renewable resources. Renewable resources can be regenerated, but the regeneration rate and the total amount are fixed. To a large extent, the ethics of ecological civilization are to maintain and improve the stock of renewable resources and their regeneration rates and outputs. These resources are associated with land, water, and biological production systems. The land area on the Earth is constant and cannot increase, but its quality or productivity could be improved or degraded. Water is renewable, but its amount at a time and space and distribution are not constant. Water and soil erosion can weaken local water conservation capacity, and the degradation of ecosystems can change the water cycle. Unrestrained renewable resource exploitation and utilization will cause the destruction of productivity and the collapse of ecosystems. Thus, food security not only depends on the amount of land but also depends on the resource-associated land quality. Land productivity with certain levels of output is not only the foundation of human existence, it also supports biodiversity and sustains the functioning of ecosystems. In Chinese history, most of the natural disasters were sudden plunges of land productivity due to extreme climate events and consequent severe food shortages. The campaigns of converting farmland back to forest, grassland, and lake in the late 1990s were efforts aiming at restoring the natural productivity.

If the natural constraint can be described as an external boundary, human biological needs are an internal constraint on the size expansion of economy. The quality of a person's needs for such necessities as food, clothing, housing, and transport can be of great difference, but the quantity of such needs should be limited. Take nutrition intake as an example. Too low daily calorie intake will cause malnutrition, while too high daily calorie intake will cause overnutrition. How much clothes a person needs to wear depends on weather conditions and the person's health conditions. In many cases, natural is the best. For example, between the options of natural ventilation and artificial air circulation system, natural ventilation is natural and more suitable for people. Under the ecological civilization, people's lifestyle choice does not necessarily need to prefer modern or artificial options.

The second aspect of population constraint is the limit of population growth. Thomas Robert Malthus's book *An Essay on the Principle of Population* or the theory of "population explosion" in the 1970s attempted to demonstrate that population multiplies geometrically and food production, due to limitation of natural productivity, grows arithmetically; the latter could never catch up with the former, and ultimately, population growth would consume all the fruits of development. From a practical perspective, geometrical growth can be hard to detect. Just like the growth of lotus leaves in a pond, if the total size of lotus leaves doubles every day, the pond will be completely covered 30 days later, suffocating all other life forms.

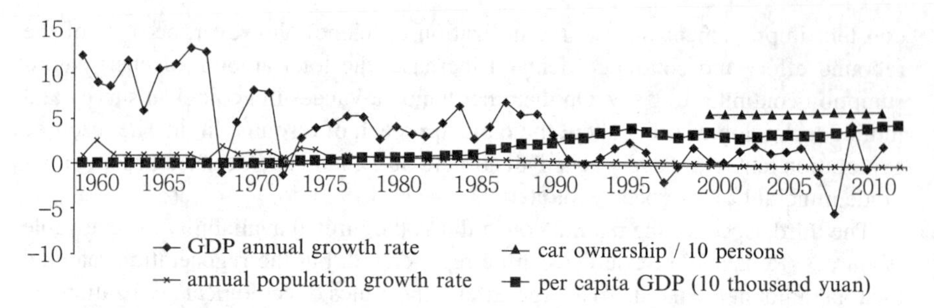

Fig. 7.1 The changes of Japan's average annual population growth rate, economic growth rate (%), car ownership (number of cars per ten person), and per capita GDP (ten thousand US dollars, at market exchange rate and current prices) from 1962 to 2013 (Source: World Bank database)

But the danger may not be noted until day 29, because on that day, the leaves cover only 50 % of the pond. However, in modern society, postindustrial society, or ecological civilization, the tragic consequence of geometrical population multiplication will not necessarily become an inevitable threat. Population in developed countries is stable, and it even declines in some postindustrial countries (see Fig. 7.1). The strict enforcement of the one-child policy since the late 1970s has decreased China's annual population growth rate from more than 2.5 % in the early 1970s to less than 0.5 % in the early 2010s. After over three-decade enforcement of the one-child policy, statistics from the Chinese Commission of Population and Family Planning, until 2011, the one-child policy covers 35.4 % of the total population in Mainland China, 53.6 % of the population are allowed to have "one and a half" children per couple, 9.7 % of the population are allowed to have two children per couple (including couples of some minority ethnic groups and couples with both husband and wife having no sibling), and only 1.3 % of the population are allowed to have three or more children (mostly herds from minority ethnic groups in Tibet and Xinjiang). Data of the sixth national census, which was carried out in November 2010, showed that the average annual population growth rate is 0.57 % from 2000 to 2010. Compared with the average annual growth rate of 1.07 % during the previous decade, from 1990 to 2000, the average annual growth rate had nearly halved. In 2013, the proportion of people over 60 in China reached 13.26 %, which means that China has become the aging society. The United Nations' population projections updated in 2010 indicate that China's population will peak in 2025 with the total number less than 1.4 billion and by 2050, China's population will decline to below 1.3 billion by 2050, and it will further decrease to 940 million by 2100. In November 2013, the Third Plenary Session of the 18th Central Committee of the Communist Party of China announced the decision to ease the population policy. According to this decision, most families will be allowed to have two children. However, the social reactions indicate that easing of the one-child policy will not affect the long-term trend of population growth in China.

Since the industrial revolution, the human capacity to create material assets has improved greatly due to technology innovation and progresses in engineering techniques. These material assets include railways, highways, ports with modern logistic facilities, airport, large public and private buildings, energy service facilities, large water supply facilities, as well as drainage and sewage treatment facilities. Traditionally and historically, most buildings and houses in China were built with wood and clay, and they were of poor quality and short use life. Therefore, as material assets, only a small number of old Chinese buildings still exist. Three famous Chinese historical buildings—Yueyang Tower, Huanghe (Yellow Crane) Tower, and Tengwang Pavilion—have been rebuilt many times in short intervals due to fire or natural wear and tear. For example, the Huanghe (Yellow Crane) Tower in Wuhan City was originally built in 223 AD during the Three Kingdom Dynasty, and since then it had been damaged and rebuilt many times; during the Ming and Qing Dynasties (from 1368 AD to 1911), it was damaged seven times and experienced ten times of rebuilding and major maintenance. Its last rebuilding during the Qing Dynasty took place in 1868, only to be flattened 16 years later in 1884. In industrial civilization society, due to progress in construction technologies, the reinforced concrete buildings could last for more than one hundred years. In October 1981, the Huanghe Tower was rebuilt again with modern engineering techniques; the tower used to have three floors, and the rebuilt one consists of five floors and has a total floor area of 3219 square meters, much larger than before. The new tower is equipped with fire- and earthquake-resistant designs and facilities, and the maintenance requirements are low. Data from the World Bank indicate that since 1980, the United States, the European Union, and other developed countries have seldom invested in extending their railways, and in some countries, the total railway length has declined. For example, in the United Kingdom, the railway mileage fell from 18,000 km in 1980 to 16, 000 km in 2012. During the same period, China's rail mileages increased from 50,000 km to 63,000 km.[11,12] It's thus clear that when the stock of material assets is close to saturation or reaches saturation, a country only needs to invest in their maintenance and renovation. Compared with building new material assets, maintenance and renovation have much small boosting effects to national economic growth. Through the reconstruction after World War II, the housing stock in European countries, which are postindustrial societies, could meet the demand or even exceed the demand. The housing stock is saturated, and as a result, it does not need large-scale investment in housing, and it is impossible for the housing prices to soar or grow beyond the working class's affordability. The stock of other durable consumer goods is also saturated. For example, in the past decade, car ownership per thousand people in the developed countries has remained stable or even declined in some cases. The car ownership per thousand people in the

[11] These countries and regions include China, India, Japan, Russia, Hungary, France, Italy, Spain, Luxembourg, Britain, Germany, the Netherlands, Israel, Saudi Arabia, Kuwait, Brazil, Angola, Afghanistan, Nigeria, Chad, Ethiopia, Ghana, Mozambique, Nicaragua, Portugal, Rwanda, the United States, Vietnam, Asia, Europe, and Africa.

[12] http://data.worldbank.org/indicator/IS.RRS.TOTL.KM.

United States fell from 473 in 2000 to 403 in 2011, which indicates that family car ownership in the United States had peaked.[13]

Since 1990, the Japanese economy has been in stagnation; thus, the 20 years after 1990 were considered to be Japan's "lost 20 years." Figure 7.1 shows the changes of Japan's average annual population growth, economic growth, per capita GDP, and number of car ownership during the 50 years from 1962. Japan's annual economic growth rate declined from around 10 % in the 1960s to approximately 5 % in the 1980s, and then it plummeted to zero or even negative growth after 1990. Such changes are apparently closely related to Japan's postwar industrial expansion, urbanization, and material asset accumulation and saturation. Japan's average annual population growth rate declined from 1 % in the 1960s to 0.3 % in the 1990s, further to negative growth since 2006. There has been no Malthusian trap of geometrical population multiplication and no sign of population explosion; instead, Japan's population shows a natural inherent limit.

The analysis above shows that under the industrial civilization development paradigm, countries have overcome the low productivity and the Malthusian trap of overpopulation, which are typical features of the agricultural civilization. Under the industrial civilization development paradigm, the material wealth became much more diversified and abundant and accumulated quickly, and the economic growth hit its natural limit. However, the high productivity is realized at huge material consumption and large pollutant emissions, which have almost reached or even exceeded threshold of the planet's environmental carry capacity. Despite the security of material supply for livelihood, effective disease control, and dramatic improvement in health conditions, population explosion has not occurred; instead, it has been stable or even declined in some industrial countries; the population has peaked in many developed countries. Once a country changes into a postindustrial society, the material wealth which human society needs and pursues will be saturated, with little space for further expansion due to the double constraints of environmental and social demands.

Thus, the growth limit under the industrial civilization development paradigm is not because of the Malthusian population explosion but the rigid physical boundary of the Earth, the peaking of population, and the saturation of the material asset stock. With these triple constraints, economic growth based on continuous industrial expansion will become impossible or unnecessary in the postindustrial society. Given that the industrial expansion growth will encounter the triple constraints, how can the economic growth continue, so to meet the needs of social and economic development?

7.3 Ecological Growth

In face of the economic growth slowdown, the government tends to hold the traditional opinion under the development paradigm of industrial civilization that "emphasizes economic growth and finds recession unacceptable," This insistence

[13] Source: WDI database.

on economic growth is a deep-rooted mind-set under the industrial civilization. However, in developed countries, which have entered the postindustrial development stage, the panacea of "stimulating economic growth" often no longer works and in some cases even backfires.

When the recessions in economic cycles occur, in order to shirk their responsibilities, the government tries to cover them up through injecting more government investments. As a result, the overall economic efficiency of the society declines, and the government's control on economic activities increases, causing a higher degree of centralization, restraints on private economy, and creativity, leading governance to the wrong path. Strong government interventions on economic activities can only work in short term, and their effects cannot last. The government uses the institutional mechanisms of industrial civilization to ensure market stability, absolute equity, security, and governance legitimacy; these efforts are actually seeking temporary stability by sacrificing the future.

The government could increase public investment due to its huge borrowing capability, but all the debt has to be paid, through either transferring them to the next generation or quantitative easing printing more money. As quantitative easing decreases the value of money, it can reduce the debt burden, but it is essentially spending tomorrow's money for today's expense. Actually, economic recession can eliminate outdated productivity through market competition, make room for new productivity, correct the imbalances and market distortions accumulated over years, and clean the way for the next round of development. Denying the existence of economic cycle and covering up the problems through borrowing money and increasing government investment are imaging indefinite high-speed economic growth, abusing public power, refusing to accept any policy failure, and choosing to cover past mistakes with an even bigger mistake. Under such a background, the decision of "quantitative easing" is made, and a lot more money is printed and injected into the economy in the name of stimulating growth. This generation, who stimulates economic growth through quantitative easing, is getting old; the next generation will grow up and become new force for economic growth. But the quantitative easing policy makes life costs soar; young people find it too expensive to have a family and choose to postpone and give up the idea of marriage and having children. The quantitative easing policy not only suppresses the development of this round but also the next round. Although the older generation finds the market value of their personal assets becoming higher, thanks to the quantitative easing, they do not increase their consumption accordingly; thus, social wealth is locked in wasteful uses.

In the 1980s, the Japanese economic strength and the economic prospects looked encouraging. Japan's purchasing power became prominent in the world, and the consumers were optimistic about their future economic growth, and some even forecasted or had the ambition of buying the United States. However, the economic bubble suddenly burst in 1990, and the dream about continued high economic growth was thrashed. In order to revitalize the economic growth, the Japanese government has been implementing a monetary policy of low or even zero interest rate since the early 1990s, but the Japanese economy has not grown much. Similarly, the United States adopted a quantitative easing policy after the financial crisis in 2008, but the effect is also not obvious. China injected a "strong" stimulus of four trillion yuan into its economy in the wake of the 2008 financial crisis; 5 years later, the Chinese economy is still struggling despite the massive increase in government investment. This means that economic growth model under industrial civilization is phased. Once across the stage of industrialization, countries have to actively seek economic transformation and find a new growth model.

Britain is the birthplace of the industrial revolution and also the first country that has accomplished industrialization and changed into a postindustrial society. Its urbanization rate is close to 80 %; hence, there is little room for relying on large-scale urbanization investment to maintain and stimulate economic growth. In Britain, the stock of infrastructure, housing, car ownership, and other assets is more or less saturated, the population is relatively stable, the domestic market is small, and its exports lack competitive advantages due to its high labor cost. Thus, the UK government has long stopped pursuing economic growth through industrial expansion that relies on high resource consumption and high inputs. During this period, Britain's forest coverage rate has kept increasing, which means that Britain allocates more land resource for natural conservation and forest construction, instead of urban expansion and industrial development. The most prominent feature of economic growth model change in Britain was economic restructuring. In 1990, the tertiary industry only contributed 66.6 % of Britain's national GDP; in 2013, its share rose to 79 %, with an average increase of 1 percentage point each year. During the same period, the secondary industry's share in its GDP had shrunk from 31.8 % to 20.3 %. The economic restructuring was almost entirely the shrinking of the second industry and the corresponding expansion of the tertiary industry (see Fig. 7.2).

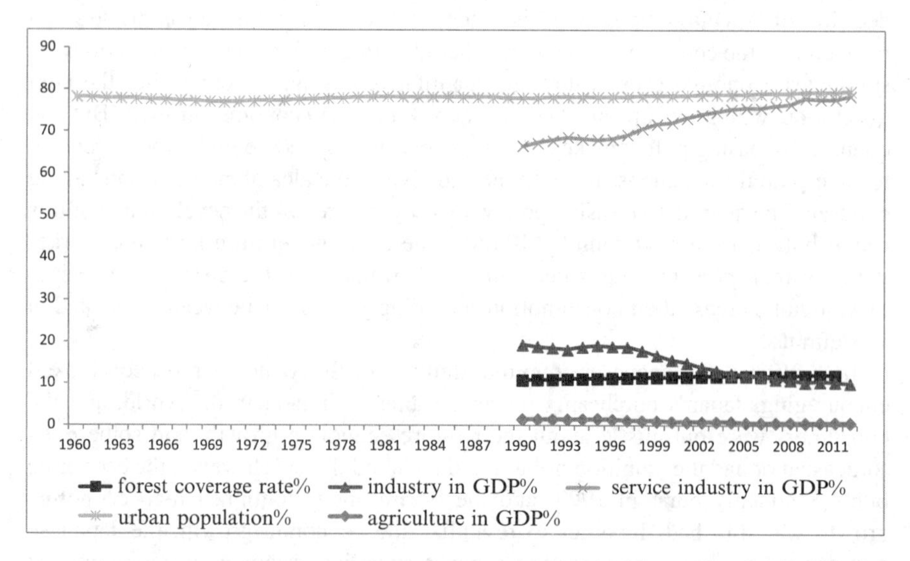

Fig. 7.2 The trend economic and social structure and environmental changes in the United Kingdom, 1960–2013 (Source: the World Bank database)

Japan entered the postindustrial stage much later than some European countries, and its urbanization process provides a good indicator for illustrating the economic and social transformation. In 1960, British urbanization level was already close to saturation, with little increase afterward, while Japan's urbanization rate was only 63.3 % in 1960, 15.1 percentage points lower than that in the United Kingdom. Japan's forest coverage rate was as high as 68.4 %. Generally, it's entirely possible for Japan to use its forest land for urban or industrial expansion to stimulate the economic growth. However, while the urbanization rate rose from 77.8 % to 92.3 %, Japan's forest coverage rate did not decrease but increased 0.2 percentage point. Similarly, the industrial structure of Japan, a developed country with strong industrial competiveness and a highly developed manufacturing sector, has also undergone tremendous changes. The proportion of service industries grew from 64.7 % to 73.2 % from 1960 to 2013, while the proportion of its secondary industries declined from 33.4 % to 25.6 % (see Fig. 7.3).

Britain and Japan accomplished urbanization and industrial restructuring while increasing their forest coverage and effectively reducing environmental pollution. To some extent, their economic development can be described as ecological transformation. Despite their low economic growth rates, the social security and welfare for their citizens has not reduced, and their per capita national income has even seen some increases. Therefore, it's an eco-friendly economic growth. What are the characteristics of ecological growth? First of all, the economic growth is slow, modest, and sometimes even negative, but this growth avoids big fluctuations in growth rate, and generally the economy development is stable and smooth. Second, the natural environment is further improved, with the natural assets constantly increasing. Japan and Britain's constant expansion of forest, which has the highest primary

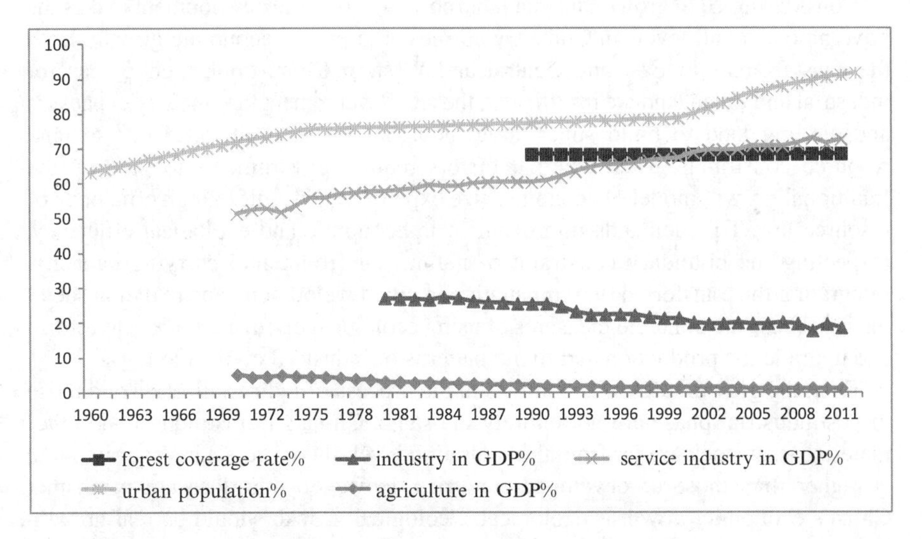

Fig. 7.3 Changes in the economic and social structures and environment in Japan, 1960–2013 (Source: World Bank database)

productivity and biodiversity among all terrestrial ecosystems, provides a good example. Third, the low or even negative economic growth and reduced industrial and urban land use do not lead to a decline in people's living quality; on the contrary, people's living standard is improved. Fourth, the triple constraints are recognized and complied. With the increase in natural asset stock, the gradual saturation of material asset stock, and the stabilization of total population, countries successfully avoid the Malthusian trap. Fifth, ecological growth must be real growth which can lead to the accumulation of social material assets. If a building is built and demolished shortly afterward and the process repeats, the GDP will grow, but it is a false, futile, and wasteful growth leading to no actual increase in social material assets. However, it should be noted that the per capita consumption of fossil energy and carbon emissions in developed countries is much higher than their shares of the available carbon emission budget for global climate protection. Therefore, at the global level, the consumption patterns in developed countries have exceeded the carrying capacity of the environment, and their economic growth is not entirely ecological growth.

China, which is still in the middle and late stage of industrialization, has two options. The first option is to complete all traditional steps of industrialization and postpone the transformation until the whole of society enters the postindustrial stage. The second option is to carry out the transformation in the process of industrialization to achieve ecological growth. As a late comer, China cannot follow the industrialization path of the developed countries due to resource and environmental constraints. On the one hand, China has to continue the process of industrialization and accelerate the accumulation of material wealth to approach the saturation level as soon as possible; on the other hand, the traditional model of industrialization should be transformed and upgraded through the building of ecological civilization.

Considering GDP growth, the main purpose of economic development makes the governments at all levels in China try all means to pursue economic growth. Both the coastal areas in East and Central and Western China think their quotas for industrial and urban land are insufficient; the city development through size expansion and circling land to build super large industrial zones is impeded due to land resource constraints. Such external factors require governments to give up the traditional growth model of economic size expansion and shift to high efficiency of resource use. This demands improvement in economic and ecological efficiency, respecting the boundary constraint of natural environmental carrying capacity, integrating the transferred rural population in city development, and realizing social justice. These are concrete measures of using ecological civilization ideas to reform and upgrade the production and living patterns of industrial civilization.

First, the growth rates of economies in the process of industrialization and those in postindustrial phase are not entirely the same. China's infrastructure and other material asset stock are far from the saturation level; therefore, China's growth rate is higher than those of developed countries. However, this does not mean that China's economic growth is ecological. Ecological growth should be real growth. Whether the economic growth is real or not can be judged from two aspects: first, whether there is a real material asset accumulation and second, the resource and environmental impacts of economic growth. Many villages, which are surrounded by new urban buildings due to city expansion, may not have high resources and environmental impacts. However, these village houses and buildings have been built in the absence of government planning and coordination and are of poor quality and lack infrastructure facilities, and they need to be demolished or rebuilt sooner or later; thus, they are not valuable social material assets. First, due to China's uneven economic development, the economic growth speeds also vary in different regions. For example, the growth rate in East China, due to the strong triple constraints, will be lower than those in Central and Western China. Second, the natural asset stock and environmental quality are improved. Due to China's fragile ecological environment and its low carrying capacity, the industrialization process should not only consider reducing environmental damage because of economic growth but also improving the environment. Third, the growth should be realized in association with improvement in social justice. If the economic growth cannot benefit ordinary people and the poor, the social income gap will widen, leading to instability, which will pose a threat to natural assets and material assets accumulated in the course of economic development. This is why in most countries, higher development level is accompanied with more equitable social income distribution. And the Gini coefficients of least developed countries and countries that have not come out of the middle-income trap are much higher than those in developed countries. This implies that ecological growth is also human-oriented growth, not growth just for the benefits of a few people. Fourth, ecological growth must be growth in harmony with nature and respecting nature. With the application of advanced technology of modern industrial civilization, skyscrapers are built in many big cities, which can stimulate economic growth and accumulate material assets. However, a lot of ecological assets need to be consumed to maintain and operate these buildings. In fact, it's not a kind of development that conforms to the nature.

7.4 Steady-State Economy

Economic growth under the framework of ecological civilization should constantly seek for ecological growth until it ultimately enters the stage of steady-state economy. The so-called steady-state economy could be achieved through two ways: active transformation and passive transformation. The ideal steady-state economy in the theory of John Stuart Mill was an active choice, while Herman Daly argued that steady-state economy was a passive response to the boundary constraints, which should be regarded as a choice in respect to nature, and the limit constraints, according Thomas Robert Malthus or the Meadows, are a passive and helpless dynamic equilibrium which, in fact, is an unsteady state.

China's agricultural economy, which lasted for thousands of years, is a steady-state economy in some aspects, but it is mainly a passive adaptation. With advancement in agricultural technology and stable social environment, the economy grew slowly; therefore, it was a dynamic steady state. However, the agricultural economy, with its low productivity, poor quality of life, and scarce material wealth, obviously is not the ideal steady-state economy that Mill looked forward to. To a greater extent, it can be regarded as a dynamic steady state with of certain properties of the Malthusian theory. Since the founding of New China, China has been vigorously promoting the industrialization. Since 1949, China has massively developed its industry, and the Chinese economy is a developing economy focusing on size expansion and undergoing continuous growth. Despite fluctuations, China's economy has been in constant growth on the whole.

After three decades of rapid growth since the reform and opening up, China has accumulated some material wealth, but there is still a large gap between China's development level and that of the developed countries. Therefore, China's expectation and desire for rapid economic growth are still very strong. Due to its poor environmental carrying capacity and the rapid consumption and decline of its natural resource stock, the traditional development model has been questioned. Chinese car ownership per thousand people was less than 10 in 2000, and now it is close to 100. China is accumulating material assets at a high speed. Although China's car ownership rate is only 1/4 than that of developed countries, China's dependence degree on oil import is more than 60 %. The congested traffic and severe smog in city indicate that car ownership in Chinese cities has already reached or approached the limit of environmental capacity. On the other hand, China's population development pattern has been fundamentally changed. The Malthusian trap, which caused much concern in the 1970s, is no longer a threat for China. Moreover, it is projected that China's population will peak in 2025, 15 years earlier than the expected peaking year predicted 10 years ago; the projected peak population size is 20 % smaller than the earlier predictions. By 2100, China's population will be about 1/3 less than the current size.

There is some conflicting situation. On the one hand, there is some space for China's material wealth accumulation. On the other hand, the rigid constraints of the natural environment are becoming more and more prominent, and the limit of the population constraint is also approaching quickly. From Japan's growth trajectory, we can see that Japan maintained a high growth rate of 10 % per year for almost 30 years from the 1950s to 1970s; then the annual growth rate declined to 5 % during the transition or adjustment period in the 1980s, and its growth rate has been near zero since the 1990s. In the process, Japan transformed from a rapid expansionary economy to a steady-state one. Japan's population growth rate in the 1980s was about 0.5 %, roughly the same as the current growth population rate in China. In the 1990s, when Japan's economic growth was near zero, its population growth rate was about 0.3 % per year; in the twenty-first century, Japan's population growth further declined and has been negative since 2006. Although China's population policy is easing to some extent and its population change may be slower than that in Japan, China's overall population development trend is to some extent similar to that of Japan. Moreover, the population is aging more rapidly in China; especially for tens of millions of families who lose their only child, their needs for material assets and consumption are small. In this sense, the Japanese economy's close to zero growth rate but high-quality growth is of some reference value to China (Fig. 7.4).

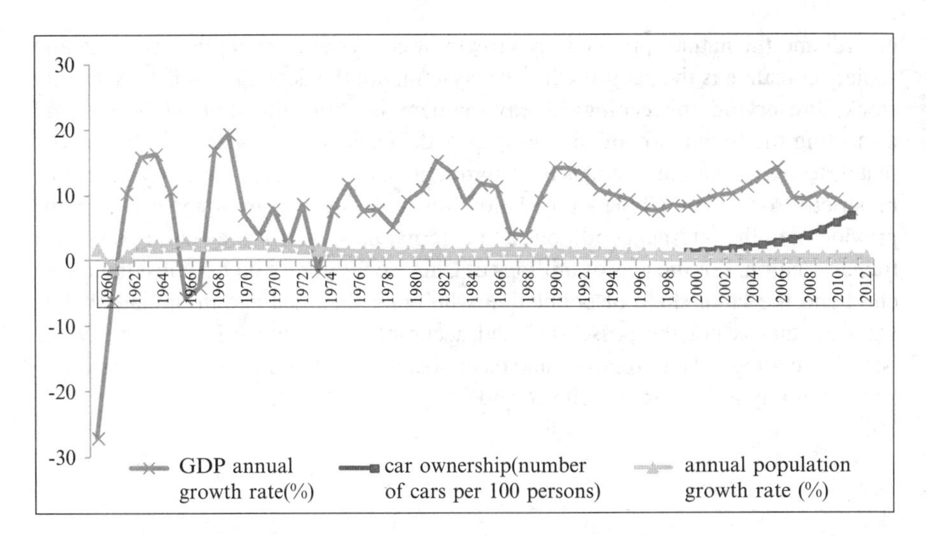

Fig. 7.4 China's economic transformation: toward a steady-state economy? (Source: World Bank database)

In such case, the slower growth of China's economy is normal or inevitable. It is a normal phenomenon that Japan's economic growth plummeted from 10 % per year in the 1970s to 5 % per year in the 1980s. Therefore, people should not be surprised if China's economic growth rate falls to 5 % per year during the 13th Five-Year Plan (2016–2020). It is also normal if China's economic growth rate further slows down to 3 % per year in the 2020s.

Given the above projections, what China faces now is not a slowdown in economy growth but the inevitable trend that China's economy will shift from high-speed growth to low-speed growth and further to near-zero growth. First, China's current priority is to emphasize the growth quality instead of growth rate, making sure that the economic growth is real and generates actual material asset accumulation. Second, or more importantly, China needs to secure social justice. Japan's economic growth transformation happened very fast, and in 10 years, its annual growth rate declined from 10 % to nearly zero. There was little social unrest in Japan during the period, mainly because that Japan's income distribution is relatively equitable. Japan's Gini coefficient is low than that in many other developed countries. China's dual system of urban and rural resident registration which originated from the historical urban–rural dual structure and the dual system of high-income state-owned economy and low-income and low-competitiveness private economy are serious obstacles to the equitable and universal distribution of economic growth benefits to all Chinese citizens. This makes China's society more vulnerable: how to maintain social stability in times of major slowdown in economic growth is a difficult task facing the government.

Third, China needs to protect its ecological environment. The near-zero growth is actually a kind of ecological growth. There is some linkage between economic growth and the natural productivity growth of ecosystems. What the economy and society consume is the net growth of ecosystem, not the depletion of natural capital stock. Preserving the ecological environment is protecting human homes and protecting the foundation of economic growth. Thus, any production or life model that destroys the nature, pollutes the environment, or endangers the ecosystem must be strictly forbidden. China's population, resource, and environmental conditions provide both the internal conditions and external pressure for the country's growth transformation. For the tens of millions of Chinese families who have lost their only child and the large share of population who enter their old age in the late 2020s, what matters is not the possession and accumulation of material wealth but the beautiful ecological environment and basic social security. Moving toward a steady-state economy is a conscious choice and an inevitable trend.

Chapter 8
Consumption Choice of Ecological Civilization

In the agricultural civilization era, the productivity was low and material goods were scarce; hence, fulfilling basic needs of human beings was the fundamental goal of consumption. In the industrial civilization era, large scale of efficient and mass production made material goods abundant, and the consumption surpassed the basic needs, and the society entered an era of mass and high consumption. Actually, such unsustainable consumption mode accelerates the termination of industrial civilization paradigm. The ecological civilization paradigm seeks sustainable consumption that is of high quality, healthy, and eco-friendly. The new paradigm is not only an ethical principle or a social choice; it also needs the frame of system regulations of ecological civilization to promote paradigm transformation.

8.1 The Natural Attributes of Consumption Choice

As human activities are a part of natural ecosystem, it is impossible and unnecessary for people's consumption to be infinite, even in the era of extremely abundant material goods. This is because, on the one hand, people's consumption is constrained by the availability of natural resources, making it impossible to be infinite; on the other hand, as a biological individual, human's material needs also have biological limitations. The rational consumption of respecting nature will be subject to natural and biological boundaries. Therefore, even if human beings do not respect nature actively, their consumption has to follow natural rules passively, from both natural physical and biological perspectives.

A basic philosophy of human's consumption culture is the desire and pursuit for goods or assets. This philosophy has the positive effects on economic development and material living standard improvement. However, it also has some irrational elements, which have negative impacts on living quality improvement and social progress. Actually, material consumption is only a part of human's life, and it is impossible to be unlimited, and any excessive consumption will have negative

J. Pan, *China's Environmental Governing and Ecological Civilization*,
China Insights, DOI 10.1007/978-3-662-47429-7_8

impacts on human society. The natural or physical constraints mainly refer to the quantity limit of natural resources. There is only one Earth and the total amount of land resource and water resource is fixed. The reserves of nonrenewable fossil fuels are fixed and cannot be increased in short geological period; once the fossil fuel resources are exhausted, it is impossible to renew them. Based on existing proved fossil fuel reserves and current world annual consumption, oil reserves can support 40 years of world consumption, while coal reserve is enough to sustain 200 years of world consumption. Although through technology innovation, the Daqing Oilfield and Shengli Oilfield of China have adopted advanced tertiary oil extraction technologies, their crude oil outputs continuously decline, their costs constantly increase, and their crude oil reserves are depleting. Even if humans could find substitutes for gasoline, automobiles need parking spaces and roads to drive on; the size of parking space and length and width of roads will be subject to rigid constraint of limited spatial surface, especially the limited space suitable for human settlements, which makes it impossible to realize the dream of everyone owning a car. Solar energy is stable and does not deplete, but the solar energy intensity in each unit area of Earth surface is fixed and changes with the circle of day and nights and the shift of seasons. Biological resources are renewable, but given the climate and technology conditions, the biological output is also limited. Confined to land territory and limited space suitable for human settlements, with the current population, China cannot realize the dream that every family has a big villa.

The technology progress under industrial civilization still works under the ecological civilization paradigm, helps alleviate the physical constraints of natural resources, and satisfies mankind's increasing consumption demand. Indeed, productivity can be increased through improving soil fertility and creating and promoting new species of crops and livestock. However, there are two inconsistencies between respecting nature and mankind's consumption demand: one is mental, and the other is physical. From the mental perspective, people can get larger living space by building more tall buildings, but the feelings of living in tall buildings, bungalows, and villas are different. When living in tall buildings, people not only need to pay for elevator operation expenses, high cost of building maintenance, and security fees but also bear the feeling of spatial compression and lack of safety. Underground space can meet human's daily housing needs, but from biological perspective, human is the product of the nature and needs green space with natural light and ventilation.

In reality, the market demand and supply indicates the human preferences for being close to nature. Market price is an explicit indicator of consumer preference. For the housing spaces of similar location, quality, and size, the market prices of basement are lower than the basements that are partially above ground; apartments with only windows facing the north, therefore unable to get sunlight, are much cheaper than apartments with windows facing the south; in tall buildings, apartments with all windows toward the same direction, therefore lacking natural ventilation, tend to be cheaper than apartments with windows toward different directions and natural ventilation. The market prices of organic, green, and natural agriculture products are higher than those produced in industrial scale. Genetically modified crops can satisfy people's biological needs for nutrition; however, due to some concerns about the safety of genetically modified products, many consumers oppose it. Although modern technology can synthesize or produce drinking water with the same contents as natural mineral water, most people still prefer natural mineral water. Such a consumption choice is a kind preference for natural products. Wastewater can be purified to meet the standard of drinking water, but tap water suppliers usually choose unpolluted water sources. By the cost of water treatment, we cannot totally explain why consumers prefer buying natural mineral water, which is much more expensive than tap water which also meets the quality requirements for drinking water. The travel consumption for being close to nature is another example of consumers' preference for natural environment; because living in a built-up environment which is far away from the nature over a long period, people mentally and physically need to find opportunities to be close to nature.

The above consumption choices with physical attributes are influenced by external factors. As living individuals and groups, human beings' consumption is influenced by internal factors. Generally speaking, the basic consumption for human development is limited, but the luxury consumption has no boundary. For example, from the perspective of utility, all kinds of transportation which can realize spatial movement within acceptable time range and comfortable conditions should be able to meet human needs for transportation. People can choose public transport, economic private automobiles, or luxury cars with large emission capacity, and the luxury consumption need is infinite. Another example is travel; it could be economic backpack travel, hedonic recreation tourism, or space tourism for adventure and novelty. Housing as a living necessity, 20 m^2 per capita can meet the living needs, but people yearn for big housing or hope to own a big one-family house with big garden. The infiniteness of human imagination and creativity determines human's infinite pursuit of material consumption.

However, as a living individual, human's consumption has a quantitative boundary. From biological aspect, the height, weight, life expectancy, and nutritional requirement of each human individual is more or less fixed. Although there are racial and genetic variations, human's life expectancy is usually between 50 and 90 years, and very few people can live over 100 years, but the natural limit of life expectancy is at most 150 years. Figure 8.1 illustrates the relationship between gross national income per capita and life expectancy. Despite the increase of income, people's life expectancy almost remains almost the same.

Figure 8.2 illustrates the situation of human's nutritional intake. In many developing countries, there exists widespread malnutrition, but in many developed countries, there is often overnutrition. Actually, in some developed countries, for example, in Germany, the per capita nutrition intake has fallen from 3600 calorie/day in the 1960s to 3200 calorie/day. Of course, different food composition has different calories; hence, there are differences in food consumption quantity and variety. Actually, such differences reflect the natural attributes of individual person's consumption choice. What people need is not a single kind of nutrition, but the combination of multiple nutrition and different food. However, regardless what food and nutrition combination a person chooses, the total amount of daily nutrition

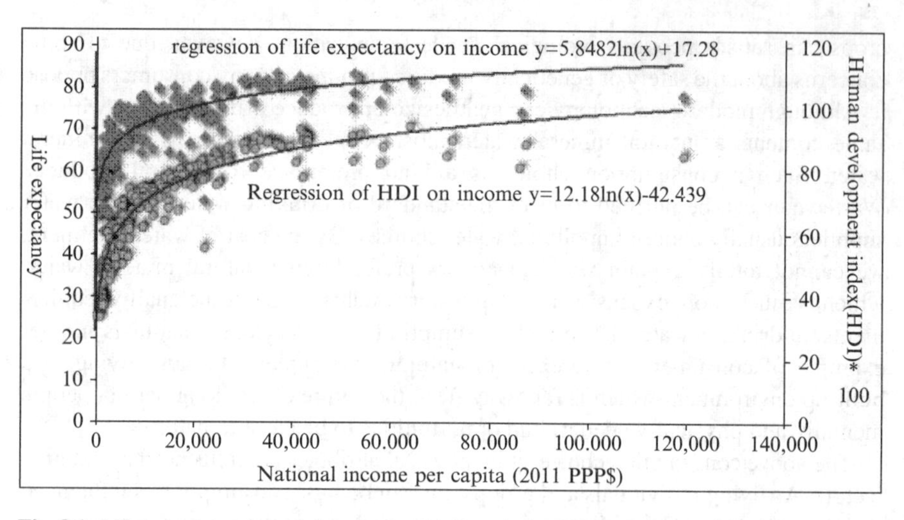

Fig. 8.1 Life expectancy (years), human development index (HDI), and gross national income per capita (PPP$/year), 2011 (Data source: UNDP (2014), Human Development Report. Oxford University Press, New York)

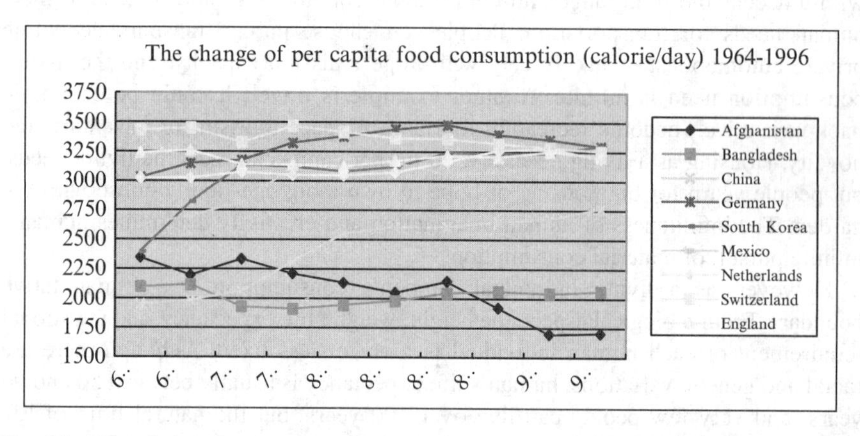

Fig. 8.2 The relations between per capita food consumption and agricultural output index (Data source: FAO Food Balance Sheets, Rome 1999)

needs of each person has a biological limit, which is an objective limit that humans cannot exceed and it is unnecessary to exceed it. The philosophy of infinite material desire and consumption is a reflection of human's possession desire, not an objective, scientific, and actual life expectancy need. As a matter of fact, in some developed countries with the aging of population, the per capita calorie daily intake from food is decreasing.[1] For example in Japan, the per capita life expectancy had increased

[1] FAO, Food and Nutrition in Numbers, 2014. Rome

from 79.2 years in 1992 to 81.6 years in 2002 and further to 82.9 years in 2014. Over the same period, the per capita calorie intake from food had decreased from 2942 calorie per day in 1992 to 2881 calorie per day in 2002 and further to 2728 calorie per day in 2014. In Kuwait, an oil exporting country, during the past 20 years, the average per capita GDP was around 80,000 $ per year, while the calorie intake from food first increased from 2144 calorie per day in 1992 to 3465 calorie per day, but in 2014 this number decreased to 3348 calorie per day. It is possible for a few members of a society to occupy a large amount of resources and realize unlimited luxury consumption, but the society as a whole has to live within the bound constraints of material consumption. In many cases, even the tiny minority's extravagant consumption is unnecessary and impossible. Recognizing and understanding the bound constraints can help form the awareness and habits of rational consumption.

There are distinctive differences between the natural attributes of consumption choice and the tendency toward high material consumption under the industrial civilization paradigm. The natural attributes of consumption indicate that human beings are a part of nature—hence, they should respect the natural constraints; as biological individuals and groups, consumers' consumption preferences also have natural attributes and are not infinite demand for various artificial goods and have a quantity limit, namely, consumption saturation. Such natural attributes are the basis for consumption choice under ecological civilization. In the era of material abundance, people's consumption is not completely physical, but a kind of natural choice. Satisfying the basic needs is to meet biological requirement with natural attribute; the consumption quantity limitation in biological sense also is a kind of natural attributes; the consumption preference for natural ecological products indicates that human beings are a part of nature. Constructing the consumption philosophy that respects nature is not only consistent with the natural attributes of consumption but also has been confirmed by market preference, indicating that the material consumption philosophy under industrial civilization is incompatible with the natural attributes of consumption and that it is necessary to establish consumption philosophy of ecological civilization.

8.2 The Consumption Value Orientation for Ecological Justice

The natural attributes of consumption reveal the limited availability of resources and the rigidity of consumption demand, that is, the guarantee of substance demand with biological basis and the infinity of material consumption desire. Because consumption choices are not mandatorily constrained by technical means and institutional design or arrangement, in order to meet rational or irrational consumption, the consumption choices in reality may not take ecological justice and social equity into consideration.

The ecological justice of consumption means that human's consumption behaviors or choices should not ignore or deprive the material consumption rights of other species and further endanger one or several species' reproduction and the function of ecosystems. The ethic foundation for ecological justice of consumption philosophy is eco-centrism[2] or bio-centrism, of which the shared value is nature orientation, not human being orientation. This ethic philosophy considers that human and other life forms each have their own intrinsic value and all kinds of life in biosphere should be entitled to equal rights, requiring to achieve equalitarianism among all species in biosphere. Human beings are just a part of biosphere, even though an important part. Compared to humans, biosphere is much more inclusive, complex, interrelated, creative, splendid, mystical, and ancient. Therefore, other species and natural resources in the biosphere are not specially designed for human's consumption and for satisfying human's demands. Human beings are an inseparable part of the natural organic and inorganic environment. Hence, human's living patterns and shared values should be adjusted and changed from human orientation to Earth orientation. From the scientific perspective, all species are resulted from the evolution process on the Earth and rely on the Earth to survive. Eco-centrism does not deny human's value, but emphasizes that human being is just a part of Earth system. Human beings' consumption should not destroy other species and natural resources that other species rely on to survive; moreover, it should leave enough survival space and consumption opportunities for other species.

The consumption ethics of eco-centrism also emphasize human's spiritual demand. The nature, including other living species and inorganic natural resources like famous mountains and rivers, is the source and basis for human's imagination and creativity, the indispensable carrier and site for soul purification, and the object and target for human cognition and knowledge progress. Once a species vanishes or natural landscapes are destroyed, the intrinsic values of the species and inorganic natural environment will also vanish, and human beings will lose the objects for physical and spiritual consumption permanently.

Only considering human's consumption demands is a deviation from the consumption ethics of eco-centrism. Unlike the consumption ethics of eco-centrism, the consumption ethics for social justice under human centrism just cares about human's consumption equity, including intergenerational and generational equity. John Rawls's social justice theory focuses on protecting the interests of weak social groups. Behind the veil of ignorance, social members make consumption decisions without clear understanding about their own social position, consumption capability, and consumption preference. According to the minimax principle, which maximizes interests of the most vulnerable social groups, the consumption choices that are in line with social justice should guarantee the consumption rights of vulnerable social groups. Take air pollution as an example; social members are unclear whether in the future they will live in places with fresh air or polluted air. In order to avoid that vulnerable social groups in disadvantage position have to live in air polluted

[2] Rowe, S. J. (1994). Ecocentrism: the Chord that Harmonizes Humans and Earth. *The Trumpeter, 11*(2), 106–107

environment, the just social choice demands that the worst air quality should meet the environment standards for human life.

In China, social equity emphasizes the equitable occupation of natural resources as a production input. From ancient time, the object and goal of social equity is to "allocate land evenly" and to enable every social member to possess the same quantity of natural resources as production inputs, but not equitable occupation and share of final consumption goods. Comparing the even occupation of production inputs, for example, land, and the equitable allocation of final consumption goods, like agricultural products, the former is more in line with ecological justice, because occupying production inputs is in order to produce consumption goods on a sustainable basis; to do that, the land users have to maintain and improve the productivity of land, respect nature, accommodate nature, and finally ensure the harmony between human and nature. The equitable share of ecological resources will lead to the equitable sharing of ecological products. If the equitable allocation of ecological products or service neglects ecological production goods, then people will be reluctant to invest in the maintenance and improvement of land productivity.

The intergenerational equity under sustainable development framework means to satisfy current generation's ecological consumption desire under the premise of not damaging future generations' ecological consumption. Because future generations cannot influence current generation's ecological consumption decisions, they are in a relatively disadvantaged position. Hence, the social justice under minimax principle of the veil of ignorance should guarantee the ecological consumption of future generations. This means that the current generation needs to protect the ecosystems and maintain biodiversity, so that the future generations could have the same right and opportunity to enjoy the basic ecological service and products. If current generation destroys the environment and consumes natural resources beyond the carrying capacity of ecosystems, they will infringe future generations' consuming rights and interests. The even allocation of land is in line with intergeneration equity. Since the production inputs are limited, intergeneration equity demands that each generation should produce and consume within the limited carrying capacity of natural resources. Such production inputs are limited both in quality and quantity. Being responsible to future generations requires that the current generation should not degrade production inputs, which are the basis for ecological services and products, in quantity and quality. In other words, the ecological assets will not decrease with time and among different generations.

The intragenerational consumption equity demands that different social groups and different individuals within the same social group should equally occupy ecological assets or equally share ecological service or products. However, such equal occupation or equal sharing could by no means be absolute equalitarianism. It can be achieved in three ways. Firstly, the basic demands of social disadvantaged groups should be satisfied and guaranteed. The harmony between people demands consumption dignity. If the basic needs for survival cannot be satisfied, then such social groups and individuals lack survival security, and the social harmony cannot be realized. Secondly, it is necessary to promote rational consumption. The harmony between human and nature does not allow consumption that plunders and destroys nature or surpasses natural carrying capacity or luxury and wasteful consumption. Thirdly, the consumption of one social group or individual should not harm that of other social groups or individuals. The consumption under industrial civilization is of high pollution, high emission, and high resource use; at the same time, while meeting a person's consumption need, it generates a huge adverse impact on consumption of other social groups and individuals, such as ground water pollution and air pollution that may affect other regions.

Ecological justice requires that the rich social groups and individuals undertake more social responsibilities. The reason is that the consumption orientation and styles of rich people influence the whole consumption orientation and styles of the whole society, while the disadvantaged social groups' consumption orientation has little impacts on the society and the future. In addition, the rich groups are able to choose the consumption patterns of high efficiency and low emissions, which can reduce ecological destruction and resource consumption. Most importantly, rich groups could afford luxury and wasteful consumption, occupy other social groups' ecological resource, and surpass the Earth's carrying capacity. For example, if rich members of the Chinese society do not spend money on purchasing luxury cars with high emissions, but buy solar photovoltaic power generation equipment to generate renewable energy and replace coal-fired electricity and gasoline, their consumption will be sustainable and in line with ecological justice. Such consumption choices can lead the whole society's transition toward consumption.

8.3 The Eco-friendly Rational Consumption

The consumption orientation under ecological justice should be equal, rational, and eco-friendly. The consumption paradigm under industrial civilization is high consumption. In order to realize high consumption, it is necessary to create more wealth and induce and stimulate ever more consumption. It forms the circle of material desire and consumption–production expansion–wealth growth–more consumption. Such consumption circle can stimulate economic growth and increase GDP and is constantly driven by the industrialization process. At the same time, this circle is also the root causes of ecological degradation, environment pollution, and resource depletion. The consumption orientation of ecological civilization paradigm should weaken or even break this circle and realize sustainable production and consumption.

The material desire and consumption focuses on income increase and material consumption. Both classical economics and neoclassical economics, which are based on utility maximization, interpret development as economic growth and further define growth as income increase. This is because material consumption brings about utility and the measurement of utility is in monetary terms. The increase of monetary value is equal to utility growth and the improvement of social welfare. The economic development stage theory formulated by American economist Walt Whitman Rostow in the early 1960s considered the development pattern of human society as a linear process with continual material consumption increase. Rostow believed that the development of human society would experience six stages of economic growth. The first stage is traditional society of agricultural civilization, in which the productivity was low, the production relied mainly on manual labor, agriculture was the most important sector, the social material wealth was limited, and the consumption level was very low. The second stage is about to take off. The industrial civilization had appeared and the traditional agriculture civilization was challenged. The competition advantage of technology innovation expanded the world market and became the driver of economic growth, and the consumption level started to improve. The third stage was the takeoff stage, in which the industrialization process started and the accumulation rates became higher, namely, the saving accounted for more than 10 % of national income; the industrial sector developed and became the leading sector; the system framework of industrial civilization was in place, such as establishing the system of protecting private property; and government agencies that could replace private capital and make huge investment were established. At this stage, industrial civilization is in the dominant position, the economy grows rapidly, the material wealth accumulates, the consumption products diversify, and the level and quality of consumption has further improved. The fourth is mature stage. Modern technologies have spread to every economic sector, and industrial production diversifies; new dominant sectors gradually replace old dominant sectors during takeoff stage. The fifth stage is the stage of high mass consumption and highly developed industrial society. The sixth stage is seeking for high living quality on the basis of high mass consumption. In short, the economic development stages are in fact a linear process in which the industrial civilization replaces agricultural civilization, the production scale and ability increases substantially, and the consumption level and capacity improve continually. Finally, human society steps into high mass consumption stage. However, it seems unimportant whether the consumption is sustainable or not and how to realize the life quality.

Such development mode taking material desire and consumption as measurement is unidirectional; it is defined as the change of material production and consumption from low level to high level and measured by the increase of monetary income. Hence, the progress of human society is measured in monetary income and targets at monetary income increase. Although when the World Bank Development Report appraises the level of sustainable development for every country, it takes into account many other factors, the core determinant factor is still each country's GDP level. Since the reform and opening up in 1978, most of China's development goals have also been in the form of per capita GDP or doubling GDP as key indicators.

For consumers, the personal target is also singular, earning a lot of money and high consumption. Indeed, high income can provide access to various consumption choices and opportunities. Today's social consumption trends are also oriented toward material desire and consumption, such as electrical appliances, housing, cars, and tourism. However, people need to ask themselves: are high income and consumption level all the contents of living quality? Imagine that a person has high income and strong purchasing power, but (a) his health conditions are bad and he can neither drive a car nor travel. Despite his high purchase power and consumption capacity, his consumption desire cannot be realized; (b) he hopes to have many hobbies, like swimming and playing tennis and the musical instruments, but he has no time for the hobbies; (c) he has many good social opinions and ideas, but he has no personal freedom, no rights for free expression, and no political power, and his opinion cannot be published, accepted, and approved; he is under enormous ideological constraints and spiritual pressure. Can such a person of high income be said to have a high quality of life? Actually, many people are very busy to pursue high income and high consumption and neglect their bad health conditions. Some people even commit such crimes as corruption, embezzlement, or fraud and robbery for high consumption. Such focusing on material desire and consumption leads to declines in living quality, instead of happiness.

Given that monetary income and material consumption cannot objectively and comprehensively reflect the quality of life, it is necessary to explore the meaning of living quality. In the early 1950s, a scholar suggested that the living quality of humans should include nutrition status, biological conditions (height and weight), life expectancy, income levels, and political and civil rights. In the middle 1980s, Amartya Sen, an economist who understood the actual consumption situation of least developed and poor countries, put forward the post-welfarism development concept. He believed that development was to enhance the ability of people to achieve various potentials. Basic nutrition, a healthy body, work opportunity, civil rights, and political freedom are the innate rights of people. In 1990, UNDP chose income level, life expectancy, and education status as the indicators for measuring the human development level of every country. When calculating and analyzing the responsibility and burden sharing of global greenhouse gas emission reduction, since the per capita carbon emissions of developed countries are several times or even tens of times of those in developing countries, some scholars from developing countries proposed to make a distinction between basic consumption and luxury consumption. They believed that the basic material consumption to fulfill the needs for a decent life is a basic right for all members in human society.[3]

Obviously, material desire and consumption cannot reflect living quality, because it covers both basic consumption and luxury consumption. According to the basic ethics and shared values of industrial civilization paradigm, people's happiness is measured by utility. Generally, material consumption generates utility and brings about the welfare increment. However, because of the natural attributes of consumption, the limited quantity of ecological environment resources, and the saturation of biological individual's material needs, the material consumption has an upper limit. Not all increases in material consumption will generate utility; in some cases material consumption generates negative utility, which has negative or offset effects on welfare level.

Consumption can generate negative or positive utility and influence living or welfare level accordingly. Positive utility means that the increase in consumption leads to improvement in life quality or increases in welfare level, while negative utility means that the increase of consumption will decrease living quality or welfare level. Take nutrition intake as an example. If the intake of nutrients is necessary for

[3] Jiahua Pan. (2008). *Carbon budget proposal, World economics and politics*

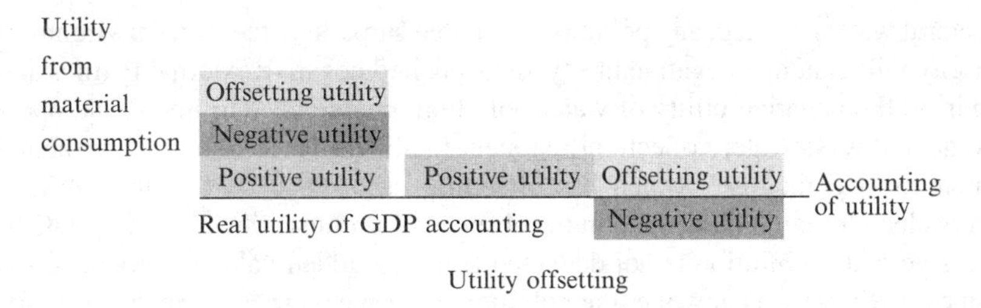

Fig. 8.3 Accounting of utility

one's growth and physical activities, such consumption has positive utility. If the nutrition intake exceeds the quantity needed for growth and daily activities and leads to accumulation of unhealthy substances in human body such as fat or adverse health conditions like high blood cholesterol concentration, hypertension, and high blood glucose, such excessive consumption has a negative utility. In order to weaken or eliminate the negative utility of excessive consumption, people need to consume more, such as a variety of weight-losing drugs and food, medicines for reducing blood fat or lowering blood pressure, and hypoglycemic and health-care products. It seems that such consumption will have positive utility, but actually it just makes people get back to normal or best living quality and welfare level. Such consumption can generate offset utility, and the total net utility from negative consumption and offset consumption will be zero (Fig. 8.3).

In the real economic and consumption activities, all consumption leads to added value and generates positive utility in economics. That means in economics, total utility is sum of absolute value of positive utility, negative utility, and offsetting utility. However, from the perspective of living quality or welfare level, only positive utility should be included, while negative utility and offsetting utility offset each other and should not be included in welfare increment. Actually lots of material consumption have negative utility or offsetting utility. One example is air purifier. In the absence of air pollution, people don't need the air purifier to purify indoor air at all. But because of air pollution and frequent smog, there is market demand for air purifiers and so manufacturers produce various air purifiers and sell them to households. These air-purifying devices just purify the air and get rid of the pollutants in the air. The polluting production activities have added value, and the production and consumption of air purifiers also have added value. But from the utility perspective, the net welfare increment is zero. This means that from the perspective of material needs and possession and consumption needs, the positive utility of living quality or welfare in biological sense is certain and cannot be increased infinitely. But the utility from economics perspective includes negative utility and offsetting utility and can be increased infinitely. The differences between utility from economics perspective and that from biological perspective will expand along with the increase of material consumption.

Because of the existence of physical limitation, there are many examples of offsetting utility; one of them is water pollution. Since the self-purification capability of natural waters is fixed, any pollution discharge surpassing the water environmental capacity will generate negative utility, the so-called negative externality. In order to eliminate the negative utility of water pollution, people need to invest and operate sewage and wastewater collection/treatment facilities. These activities are actually similar to "weight lose" which has offsetting utility, and they cannot make the waters cleaner than natural state in the absence of pollution. In economic statistics, excessive water pollution is not deducted from the added value of production and business activities; meanwhile, the pollution control expenses are included in fixed asset investment and added value. Another example is restoration and reconstruction after ecological system damages and the rescue and protection of endangered species; they actually have offsetting utility. Such economic activities cannot generate or expand new species or increase the population size of existing biological communities, but relative to the natural status of biological species and communities, they are just a kind of protection and restoration for biological species and do not really lead to social welfare increment.

Of course, the social economic activities of human beings obviously have some "normal" or "inevitable" negative utility and need production and consumption on the basis of national economic system to maintain their normal functioning. For example, people eat grains and it is not impossible to avoid diseases; the temperature fluctuations due to changes of seasons often make people catch cold. It is also normal that after middle age people's health conditions deteriorate with aging. The related medical and health consumption apparently is not due to negative utility, but a kind of positive utility. Similarly, some natural extreme climatic events and geological disasters are not caused by negative impacts of human activities. The consumption and demand for air conditioners during summer heat waves and the consumption and demand for heating supply in freezing winter and the repair of living facilities damaged in heavy storms all have positive utility. However, if some consumption activities exceed the proper limit, for example, too much cooling from air-conditioning making people sick is negative utility. The treatment for "air-conditioning disease" is to offset the negative impact of excessive consumption of air-conditioning and has offsetting utility.

Excessive nutrition intake, pollutant emissions exceeding environmental capacity, the restoration and reconstruction after ecological system damages, and the rescue and protection of endangered species are all directly related to the material desire and consumption philosophy under industrial civilization paradigm. Understanding the negative utility and offsetting utility in economic and consumption activities will undoubtedly have a positive effect on consumption preference transformation and the transition toward rational consumption. The consumption choices under ecological civilization paradigm respect the natural attributes of consumption and the natural quantity "threshold." For some products that harm societies and human health under industrial civilization, the social systems have established some rules that ban drugs and restrict tobacco demand and consumption. For some other consumer goods, for instance, wine, if people consume only a small amount occasionally, the utility is positive. However, alcohol addiction and abuse will obviously have negative impacts on society and people's health. Eco-friendly consumption seeks for rational, healthy, and quality consumption, complies with the natural limitation or carrying capacity, and protects the harmony between human and nature and between individuals and society, so as to achieve a sustainable consumption paradigm.

8.4 The Policies for Ecological Civilization Consumption

The industrial civilization is a social progress in relation to agricultural civilization, because the great improvement in social productivity and large increase of material wealth meet the material demands of human society and improve people's living standards. Therefore, the material desire and consumption philosophy is not completely bad; instead it also has rational contents. Economic growth and income increase are important factors of social progress and also a basic element of human development rights. In order to make the material desire and consumption rational, two conditions have to be met. The first is to pay more attention to living quality improvement, rather than simply focus on income levels, and especially emphasize the multiple dimensions of living quality, such as nutrition, health, education, and civil and political rights. The final purpose of monetary income increase and improvement of material consumption is to improve living quality. The second is to understand the limitation of material consumption, including the quantity limitation from biological and physical perspectives. This amount limitation requires that human society should have some rules and restrictions on material consumption. The basic material consumption and various contents of living quality are the basic rights for societies and biological individuals, which should be respected and need to be guaranteed. On the other hand, luxury consumption usually generates negative utility which needs to be neutralized with offsetting utility. As it has no positive impacts on social progress and welfare improvement, it should be restricted or banned. Especially when there are conflicts between basic consumption needs satisfaction and the physical limitation of environmental resources, effective measures need to be taken to suppress excessive material desire and consumption.

The first rational process requires human being to realize the transition from focusing on the quantity of material consumption to pursuing living quality improvement and avoid the irrational pursuit for income increase and material consumption. The other rationality requires the protection of basic human development rights and compliance with the limitation constraints of material consumption. The two kinds of rationality require the change of consumption philosophy and the improvement of environmental cultural awareness and contain many elements of environmental ethics. But on the other hand, the mandatory institutional setup and regulations can help strengthen environmental ethics and philosophy. In Europe, earning additional income through extending working hours is considered as unlawful competition and illegal—once found, such behavior is subject to economic penalty; moreover, people have to pay progressive tax for the extra income. The comprehensive social security system of developed countries can guarantee the basic needs satisfaction and various civil and political rights of all social members. At the international level, there are many conventions that strictly ban the consumption of endangered wildlife and protect their habitat, such as the Convention on Biological Diversity, the Convention on Wetlands Protection, and International Trade Convention on Endangered Wide Animals and Plants. Many countries have formulated laws and regulations to restrict or forbid products or materials that are harmful to human health and ecological system function.

In addition to the compulsory laws and regulations, various market instruments are also used to influence consumption preference and to promote the ecological justice of consumption. Many countries levy high taxes on luxury products or consumer goods which has negative impacts in order to reduce luxury consumption and raise fund, such as the high tax rates on tobacco and whisky. For the consumption of limited natural resources, in order to encourage conservation, reduce waste, guarantee basic needs satisfaction, and narrow the income gaps, levying resource consumption tax can guarantee its positive utility, curb the negative utility, and reduce the needs for offsetting utility. The examples include energy tax, carbon tax, and high tax rates for tobacco and wine in many EU countries. In Switzerland the gasoline consumption entirely depends on importation, which is almost completely consumed by automobiles, and gasoline is an important source of local carbon emissions. The Swiss government levies consumption tax on gasoline, and the tax revenue is not for government spending; instead, all gasoline consumption tax collected is allocated to every citizen according to the absolute average principle. Those consuming much more gasoline and driving cars more frequently will pay more gasoline tax, while those who do not drive private cars pay no tax and additionally can get an average allocation. This is the equivalent of compensation of high-consumption consumers to low-consumption consumers or compensation of rich people to poor people, which can be understood as a kind of ecological compensation.

Some developing countries, such as South Africa and India, also have policies to guarantee basic consumption rights and curb luxury consumption. The South African government provides 60 KWh of electricity free of charge to each poor household to ensure that they have electricity for basic lighting. South Africa lacks water, but in this country the water for drinking and cooking is almost free. However, the more water a household consumes, the higher the rates of water fees. This means that the incremental price system of resource consumption is used as an effective policy to regulate consumption, promote rights protection, and curb excessive consumption. For some basic consumer goods, such as water, energy, and housing, if the consumption is lower than basic survival requirements, they should be supplied free of charge. With the increase of consumed amounts, the fee rates should be raised progressively. Such progressive fees for resource consumption can produce threefold effect: (a) guarantee the basic consumption rights of disadvantaged social groups; (b) curb consumption of negative utility and offsetting utility and reflect limitation constraints of biology and physics; (c) raise fund for research and development and improve efficiency of resource utilization or improve the environment.

The progressive tax rate of resource consumption can be based on amount or price. It is suitable to adopt progressive tax rate based on the amount of water, electricity, and gas consumption. For some other consumer products, the price difference is very large, such as tobacco, wine, housing, health products, and automobiles, so it is better to adopt progressive tax rate based on price. Take housing as example. The value difference of houses of similar size in different locations with different building materials and quality can be several times or even dozens of times. Obviously, it is appropriate to levy consumption tax based on the market value of different houses and apartments.

Chapter 9
Ecological Institution Innovation

When a society transits to a new type of civilization, institutional innovation is indispensable to constantly regulate and accelerate the transition process. During transition from farming civilization to industrial civilization, through institutional innovation, capitalism was created and improved. While the ecological civilization institutional system will not totally deny the industrial civilization, it will be formed and improved on the basis of industrial civilization and through continuous innovation.

9.1 Motivations for Institutional Innovation

The institutional arrangements of industrial civilization cannot be perfectly in line with the requirements of ecological civilization. Countries need to deepen their understanding and include the factors of ecological civilization to reform and to improve their existing institutional system. At the 16th CPC National Congress held in 2002, ecological civilization construction was first time decided as China's national development strategy. The theories about ecological civilization construction have been steadily deepened and improved and reflected in relevant legislation and policies. China has established a comprehensive set of legislation, standards, policies, and planning system on energy efficiency and carbon dioxide emission reduction, circular economy, ecological protection, and reactions to climate change. These legislations, standards, policies, and government plans help implement ecological civilization construction in practice. In June 2002, the Standing Committee of the National People's Congress passed *Cleaner Production Promotion Law*, which was revised in February 2012. In 2005, the State Council issued "Several Opinions on Accelerating the Development of Circular Economy." Three years later, in 2008, the National People's Congress passed the *Law on Circular Economy*. Just in 5 years from 2006 to 2010, more than one hundred national and local legislations and environmental standards related to ecological civilization were enacted. Since then, a series of policies on ecological protection have been issued, including

J. Pan, *China's Environmental Governing and Ecological Civilization*, China Insights, DOI 10.1007/978-3-662-47429-7_9

"the Framework Plan on National Key Ecological Protection Area" and "the National Plan on Ecological Function Zones."

Among the 23 targets in the 11th Five-Year Plan (2006–2010), 8 are binding targets on resource and environment, accounting for 35 % of the total targets. The 12th Five-Year Plan (2011–2015) includes 12 resource and environment-related targets, of which 11 are binding ones, accounting for 43 % of the total numbers of targets. During the period of 12th Five-Year Plan (2011–2015), China's pollutant emission reduction targets include four aspects, chemical oxygen demand (COD), sulfur dioxide, ammonia, and nitrogen dioxide. The last two aspects are included for the first time. Moreover, the areas subject to pollutant emission control have expanded from industrial process and urban areas to transportation sectors and rural areas as well. A comprehensive institutional system of laws, standards, regulations, policies, and government plans on environmental and ecological protection has taken shape. This provides strong driving force and effective safeguard for ecological civilization construction in practice.

However, we should clearly recognize that some contents of the current institutional setup and administrative arrangement contain some barriers, which may impede further scientific development. China's existing legislation, institutional arrangement, and mechanisms are not completely in line with the need of ecological civilization construction. Those barriers will cause the imbalances, inharmony, and unsustainability in the process of development more prominent. Under the current ecological environmental protection system, because of institutional arrangement, policies, basic mechanisms, and some other complex factors, the efficiency and economic returns of ecological environmental protection are relatively low, and it is hard to curb the overall trends of ecological environmental degradation. This is because of problems in the following five aspects:

Split of Power and Responsibilities Currently, in most parts of China, the staff, personnel administration, and budget of local environmental protection agencies are controlled by local governments. As their hiring, promotion, and funding are in the hand of local governments, during their environmental monitoring and law enforcement, local environmental protection officials often have to listen to local governments. Although local governments are responsible for both economic development and environmental protection within their jurisdiction, local economic growth is the main criterion to assess their work performance. Economic growth directly influences whether local governments could get further resources and supports and whether local leaders could get promotion. Therefore, most local governments attach great importance to local economy development and choose to sacrifice local environmental benefits for local economic growth when environmental protection is in conflict with economic growth. Strict environmental monitoring and law enforcement by local environmental agencies increase the environmental protection costs of local enterprises, which can help improve the quality of investment, but make it harder for the local government to attract investment in local economy and lead to slower investment growth. When local environmental agencies fine polluting enterprises or order them to stop production, the local economic growth and government fiscal income may be affected and the local governments' economic interests may be indirectly hurt.

Under such circumstances, local governments often step out to hamper environmental law enforcement by local environmental authorities and protect polluting enterprises from punishment and closure. Therefore, under the existing institutional setup, as the personnel and budget administration of local environmental authorities are controlled by local government, this makes local environmental authorities unable to conduct local environmental monitoring and enforce environmental law independently.

Lack of Effective Coordination Air pollution and big river water pollution cross the boundary of local jurisdictions. Environmental protection agencies are attached to local governments, which makes it very hard for the local environmental authorities of different areas to coordinate and regulate cross-jurisdiction pollution issues in terms of pollution prevention, control, governance, compensation, and accountability. Currently, most of the inter-regional environmental coordination in China is "afterward coordination"; in other words, regional coordination is mainly temporary emergency arrangement and happens after cross-jurisdiction environmental pollution accidents and disputes occur. There is a lack of institutional mechanism for beforehand inter-regional coordination of environmental pollution provision and control. Besides, due to the power and responsibility split mentioned above, local environmental authorities are not in the position to coordinate other government departments; hence, their power is further weakened. Cross boundary issues, including provincial and city provisional cross boundary pollutants, need cross boundary institutional arrangement and working mechanisms. For example, water pollution in the downstream of the Yangtze River was caused by industrial pollution, agricultural pollution, and sewage of the provinces along the Yangtze River, as well as poisonous and harmful materials from river transport. This involves multiple provinces and many pollution sources. It is very important to find out how much pollution each province and emitter has caused in order to define their shares of burden and responsibilities. Smog in the Beijing–Tianjin–Hebei region, Northwest China, and downstream of the Yangtze River is caused by industries, transportation, and household pollutant emissions. To find out the responsibilities of each emitter and province and hold the polluters accountable, it is necessary to find out how much each entity has contributed to the smog. In the face of cross-jurisdiction environmental pollutions, the environmental governance that is based on administrative jurisdiction fails. When smog occurs, most local governments resort to restricting the use of government cars and private vehicles, temporarily limiting the productions and emissions of industrial and mining enterprises, and demanding construction sites to stop their operations. These measures could only temporarily solve the pollution problems. To solve these problems once and for all, effective and efficient institutional arrangement is urgently needed.

Besides, the wide coverage and systematic approach of environmental protection determines that most activities and entities within the administration scope of environmental authorities are also administrated by other government authorities and agencies. The complexity of environmental protection work also requires the environmental authorities to closely cooperate with other ministries and departments, including the National Development and Reform Commission, and national authorities on industries, land and other resources, agriculture, forestry, water resource, transportation, oceanography, construction, sciences and technology, security, foreign affairs, and intellectual properties.

Relying on Administrative Measures China's ecological environmental protection work mostly depends on administrative measures rather than market instruments. Charging for sewage and other emission fees is one of the most widely used measures by the environmental protection authorities to control pollution and to protect ecological environment. The pollutant emission fees are included in fiscal revenue and managed through special fund for environmental protection. This special fund is mainly used for preventing and controlling main pollution sources, regional pollution prevention, the research, development, piloting and adoption of new technologies and industrial processes for pollution control, as well as other pollution prevention and control projects decided by the State Council. Theoretically, the pollution fee actually requires the polluting enterprises to pay the environmental cost due to the environmental externalities of the pollutant emissions during their production activities. All the pollution fees are included in government fiscal revenue. This requires the finance authorities and the environmental protection authorities to effectively use this part of fiscal revenue for environmental benefits, so as to eliminate the environmental externalities due to enterprise pollution. Under the current institutional arrangement, the using process of pollution fees contains the following problems. First of all, the government agencies responsible for fiscal budget preparation and management do not have sufficient information on the actual funding needs of environmental pollution prevention and control; the research, development, demonstration, and dissemination of environmental technologies; and environmental protection project construction. Secondly, the environmental protection authorities in charge of fund allocation are unable to timely and accurately control and adjust the money allocation according to actual circumstances. Thirdly, the existing pollution fee is too little compared with the funding needs for pollution prevention and environmental restoration. Fourthly, the charge and use of the pollution fees are not transparent and public supervision is absent. These problems seriously limit the environmental benefits from the use of pollution fees.

Lacking Ecological Threshold Redlines To define the ecological boundary line is to safeguard the bottom line for ecological and environmental degradation. However, it is a complex systematical project to fix the ecological threshold redlines. It is a very difficult process, from winning support on the idea to carrying it out in practice. A series of challenges have to be overcome, including coordination between economic development targets and environmental protection targets, alignment between national interests and local interests, and solving the conflicts of responsibilities, rights, and interests among different regions and different government agencies. Mechanisms for interest coordination and ecological compensation need to be built step by step. The institutional arrangements and mechanisms for ecological boundary protection need to be improved, in order to fix the combined redline threshold for woodlands, forests, wetlands, desert vegetation, species, water resources, ocean resources, and permanent basic farmlands. Moreover, to fix the redline threshold, it is also necessary to carry out nationwide in-depth investigation and study and adopt scientific methodologies to balance between economic growth and environmental protection.

To draw ecological redline threshold, apart from fixing the minimum threshold for ecological space and resource stock, it is also necessary to gradually fix the upper limits for pollutant emission in places where conditions are mature. If there is no upper limit for total emissions in an area, then total emissions of many small emitters, whose emissions are each within the limits of relevant environmental standards and laws, may continuously grow and make the local environmental pollution higher than national standards. For example, currently environmental protection authorities at all levels focus their monitoring and administration on big polluting entities and sources. However, they do not have effective measures and policies to control and prevent the emissions from numerous small emitters whose emissions are in line with relevant regulations and standards and legal. There exist some administration gaps and blind areas. Although these small legal emission sources are each very small, due to their large amount, their total emissions are considerable and the relevant environmental impacts cannot be neglected. Take Beijing as an example, during the period from 2002 to 2012, the city's pollution fee income decreased a lot, from 192 million yuan in 2002 to 34 million yuan in 2012.[1] To some extent, it indicated that the illegal emissions, or emissions that exceed relevant regulations and standards, had dramatically declined. However, the frequent smog and the bad air pollution in downtown Beijing showed that the total emissions of air pollutants were high. A large part of the emissions are small legal emissions. Among all these legal emissions, one major source is the emissions from the millions of vehicles which meet relevant emission standards. Besides, there are also many legal emissions from industry and the construction sector. These legal emissions are out of the environmental supervision and law enforcement radar of environmental protection authorities. Their environmental externalities are left for the self-purification function of environmental system. The government will not take action to solve the problem until the pollutions far exceed environmental capacity and cause serious environmental damages. By then, it will make some big investment in some cleaning up projects. Even if the emission standards are tightened steadily, the total amount of legal emissions will still be huge and should not be neglected. Further studies are needed on how to prevent and curb these legal emissions.

The Absence of Law Enforcement The absence of environmental law enforcement is reflected in many aspects, which includes the following four:

[1] Table of pollution emission fees in local areas, 2000–2012, MEP, P.R. China: http://hjj.mep.gov.cn/ pwsf/gzdt/201312/P020131203550138828737.pdf

1. Law enforcement on rural ecological environmental protection is seriously absent. Currently, at the level of township government, there is a lack of institutional arrangement and human force for environmental protection. In some places, there are environmental protection liaison persons, team, or group for environmental monitoring and inspection. But in general, there is no uniform and standard environmental institutional setup in the vast rural areas. The human force for environmental protection is insufficient. The professional competence and qualifications of existing environmental protection personnel in rural areas are not good enough for environmental protection and monitoring. And the existing environmental protection personnel are of high turnover rate in rural areas and rural environmental protection work is rather weak. Besides, in recent years, rural economic development has been very fast, and township enterprises have been growing rapidly in terms of quantity and scale. Many industries and production capacities that are phased out in developed areas and urban places have been moved to rural areas. Township governments are also eager to attract investment in industries and construction on rural land. Moreover, environmental pollution and ecological damages are increasing in rural areas due to outdated technologies for environmental protection, weak environmental monitoring and law enforcement, and rampant secret illegal emissions by enterprises. All these factors cause pollution threats to rural environment. Rural households' living wastewater discharge, living garbage dumping, pollutions of chemical fertilizer and pesticides, and other environmental problems are not subject to unified and effective regulation under the current environmental protection system. The local environmental protection force is too weak, far from meeting the environmental protection needs of rural areas.
2. Lack of regulations on environmental law enforcement. A common problem is that in the absence of upper limits on total pollutant emissions, enterprises are charged of pollution fees based on the quantity of their emissions. Meanwhile the environmental authorities are inactive in emission control or even hope for more emissions from enterprises, so that they could get more emission fees. Inaction or action at random in law enforcement is a common problem in many functional departments of environmental protection authorities. Some environmental protection officials do not strictly enforce laws so as to effectively prevent pollution from sources; moreover, they abuse their power and force enterprises for bribes or other personal benefits. These problems reflect the lack of an effective constraint mechanism and the need for improvement on personnel education, supervision, discipline, and accountability inside environmental protection authorities and agencies.
3. Weak basic conditions of environmental administration impede law enforcement. Environmental legislation, equipment and scientific research are the basic conditions supporting environmental administration. Take soil pollution as an example, serious soil pollution is due to the long-term problem of negligence in soil environmental protection and delays in land protection legislation. In the process of environmental monitoring and law enforcement, a common problem is the difficulty to gather evidences. Many industrial and mining enterprises secretly emit air pollutants and wastewater and dump solid wastes during nights.

Without advanced monitoring devices, it is very hard to get the evidence and punish enterprises for their emissions, especially in the case of air pollutant emissions, as wind often blows the pollutants away, making it hard to trace the source of emissions. Besides, the causal relationship between pollutant emissions and environmental pollution is very complicated. It is hard to find out the causes and impacts of environmental pollution. Hence it is urgently needed to enhance relevant study and research. Many studies have tried to find out the contents, sources, and their contributions to the frequent smog in the city cluster of Beijing–Tianjin–Hebei region. Although there are some authoritative findings, most of the studies are limited to status analysis. They fail to trace the sources of pollutant emissions and estimate the impacts and extent of the emissions and environmental externalities; such studies cannot provide accurate information and data basis for policy making and designing of cost-effective measures and permanent solutions. Moreover, sometimes local environmental pollution causes serious harms to the health of villagers living nearby. However, it is very difficult to prove the causal relationship between the pollution and the health problems of local villagers; it is even more difficult for the victims to provide scientific evidences that their diseases are because of the pollutant emissions from one or more sources. It is difficult to hold the polluters responsible for the damages they cause and compensate the victims of environmental accidents.

Scientific studies are needed to overcome these barriers; they are also necessary conditions for enhancing environmental protection administration. However, in China the environmental scientific studies are weak and the government investment in them is insufficient. In the 2013 budget of the Ministry of Environmental Protection, the budget for environmental sciences and technologies was 1,524,099,800 yuan, of which the budget for basic studies was six million yuan, merely 0.39 % of the total.[2]

4. Lack of an effective system for environmental pollutions reporting and disclosure to support relevant law enforcement. In China, many cases of environmental pollution continue for a long time without any government intervention and enterprises often secretly and illegally emit pollutions. Often the local environmental authorities and local governments do not take things seriously until the illegal pollutions cause some serious harm to the neighboring areas. That reveals flaws in the existing environmental pollution monitoring and reporting system. The channels and process for pollution reporting are neither unobstructed nor convenient. The systems and mechanisms for smooth and convenient pollution reporting system need to be set up for public supervision. Besides, databases including data and information on environmental pollution, environmental capacity, environmental industries, environmental technologies, and environmental protection budget expense need to be set up and made readily accessible by the public. In this way, the public can access the information and effectively participate in environmental pollution supervision, which will greatly benefit environmental administration.

[2]Annual budget of the Ministry of Environment Protection, 2013, MEP, P.R. China: http://www.mep.gov.cn/zwgk/czzj/

9.2 System of Ecological Boundary Lines

Boundary lines are generally boundaries between different land uses. Sometimes, the redline indicating the exact location of a building along the street is called its boundary line or boundary line of building. The boundary line of building can overlap that of the street or leave some gap between the two lines. Usually building boundary line is forbidden to exceed the street boundary line. Beyond the boundary line of building, no architecture is allowed to be built. Due to the rigidity of boundary lines, this concept is now widely used to illustrate the mandatory bottom lines in government policies, for instance, the boundary line of farmlands. Ecological boundary lines are actually the boundary line for ecological security. They are the legally binding boundaries set by the government for ecological protection, in order to safeguard the normal functions and ecological service provision of ecosystems.

To respect the nature, the fundamental action is to set aside some land for keeping its natural state and protect it from the disturbance of human activities on the basis of understanding the natural space and capacity so that the ecological systems can be undisturbed and their ecological functions maintained. The design, decision, and implementation of the key ecological functional zones come from the idea of building boundary. To maintain the ecological functions of some natural ecological systems, some natural land must be earmarked to forbid development. For example, some areas are ecological barriers for other areas, can provide important ecological services, and have very sensitive or vulnerable ecological environment; these areas are designated as different ecological protection zones. Within these zones, industrialization and urbanization are prohibited to effectively protect national rare, endangered, and typical animal and plant species and ecological systems. The overall target is to maintain the main functions of important national ecosystems.

Boundary line alone cannot safeguard ecological security. The air is dynamic and water is circular. These factors' security goes beyond local jurisdiction boundaries and is subject to direct or indirect impacts of industrialization and urbanization. The security of these ecological factors is crucial for natural ecological system and social economy, especially for human survival. Industrialization and urbanization are influencing every aspect of natural system; hence, the quality of ecological factors need to be guaranteed, to make sure the public can breathe fresh air, drink clean water, and eat safe food. Therefore, the boundaries of ecological factors are not limited quantity; more importantly, they are related to certain quality standards, including air quality, water quality, and soil quality, so as to guarantee the safety and health of ecosystems and human being. Ecological factors are important components of ecological system and have some self-circulation and self-purification functions. However, large-scale industrial production and huge numbers of people gathered in the small land area of cities lead to large quantity and concentrated pollutant emissions, which exceeds the self-purification capacity of ecological factors and leads to their quality degradation. To protect the quality of ecological factors, the total pollutant emissions must be lower than their natural purification capacity. The total pollutant emissions must be effectively controlled and reduced. Risks to ecological system should also be managed and controlled. That is to say, the quality boundaries of ecological factors include environmental quality standards, caps for pollutant emissions, and environmental risk management.

Ecological systems have inherent material producing capability. Human survival and development rely on the material production of ecosystems. If the materials that human beings exploit the natural production of ecosystems, the natural productivity of ecosystems excessively will degrade. Therefore, the natural productivity of the ecosystems is another aspect of ecological security boundaries, which is ecological material consumption boundary. The total quantity of ecosystem output materials used in human social and economic development must be less than the natural productivity of ecosystems. Mankind can use technologies and economic instruments to promote the conservation of resources and energy and to improve utilization efficiency of energy, water, and land. However, the ceilings for all material uses are the quantities the nature can reproduce. The boundary line for water consumption is the quantity of water produced from the interactions among water circulation, landscape, and land topography. The boundary lines for land use aim to optimize national land development, promote the orderly use and protection of land resources, and effectively protect such natural resources as farmland, forest land, grass land, and wetland. Boundary line for energy use is the energy use level under certain economic and social development, including total energy consumption, structure, and national GDP energy intensity.

The implementation of ecological boundary in areas with high population density, especially in cities, is of particular and typical significance. That is because in cities, large population concentrate in a small space, economic activities are highly intensive, and the energy and material consumptions are high. All these factors lead to high pressure to the carrying capacity of resources and environment and high level of information and material flow. Due to lacking awareness of ecological boundaries, many local governments ignore the rigid constraint of ecological capacity. Large cities constantly build high buildings; some of them even blindly support the construction of super high buildings. That leads to deterioration of the human settlement environment. Smog regularly covers metropolitans and the urban residents are exhausted from the high living pressure and congestion in cities. Although the people's income level is growing, their losses in health and environment are increasing faster. Blue sky and clean water are disappearing in many parts of China; with the loss of these basic living conditions, the living quality of the people and the future of the country would be also lost.

Ecological boundary lines are rigid constraints. Generally, it is believed that cities have scale economy effects and they get resources from rural areas and technology progress can continuously relax the environmental constraints. Therefore, people think that the frontiers of cities can be pushed outward constantly, the scales of the cities can keep growing, and the ecological boundary lines do not necessarily mean rigid constraints for city development. However, the "urban diseases" of extreme large cities are worsening, indicating the existence and tightening of rigid ecological boundary lines. Cities can have economies of scale, but they can also have diseconomies of scale. For example, in terms of the height of a building, it is not the higher the better resource efficiency. Comparing to one-storey houses, multi-storey buildings can more effectively utilize the limited space and reduce the land use needed for building. However, taller buildings need stronger structure and the fire risk management costs are also higher. Exceeding certain heights, the operation costs grow much faster than the building height increase. The outward expansion of cities inevitably leads to declines in the economies of scale in urban transport. Besides, many Chinese cities lack water. To solve their water supply problem, cities have two options. One option is diverting water from other places and the other one is improving the efficiency of water use. Water diversion is to transfer water resource from one place to another; it cannot increase the total quantity of available water resources. Technology innovation can improve water use efficiency, but it takes time and costs money. If this efficiency can't be improved by technology innovation fast enough to keep up with the growth rate of water demand, cities' water demand will continue. In such situations, city expansion will reach or even break the boundary line of ecological security. Meanwhile, technology can generate positive and negative effects. It can support resource utilization efficiency improvement, but it can also speed up the depletion of resources. For example, when people in arid areas use advanced drilling technology to dig deeper wells for water, it helps accelerate the depletion of underground water. More importantly, at certain time and space, technology has its limitations. Human beings, animals, and plants all need water to maintain for their survival and the need is rigid. Therefore, to some extent, sometimes the ecological boundary lines are inevitably rigid. They are the boundary for city expansion and the foundation and targets for city governance.

Scientifically Understand and Set the Ecological Boundary Lines The decisions of the Third Plenary Session of the 16th Central Committee of the Communist Party of China in 2003 clearly made the requirements for setting ecological boundary lines. How to scientifically fix the ecological boundary lines? The first aspect is to assess the total boundary lines and absolute caps for each ecological factor. Industry restructuring can improve the productivity of industries under given environmental capacity; however, it cannot lead to increase the absolute amount of environmental capacity. The natural environmental capacity of air and water is fixed. Therefore it is necessary to check and estimate the inherent natural environmental capacity. For instance, the total quantity of water resources in an area is the sum of underground water volume and surface water volume; meanwhile, it is also necessary to take into account the water inflow and outflow. How much pollutants the local air could absorb is also fixed. Otherwise, smog will never occur. The second aspect is to evaluate the spatial boundary lines. Natural protection zones, water source protection zones, and urban green land have definite spatial scopes, which makes it easy to draw their boundary lines. The third aspect is to evaluate the speed boundary lines, which are the average resource consumption or pollution emission of each unit of production or unit area or each person. Unlike the total quantity and spatial boundary lines, the speed boundary lines are changing and adjustable, but they are subject to the restrictions of total quantity constraints and technology level. To set the boundary line for city expansion, it is necessary to appraise and fix a city's special boundary lines and adjust the speed boundary lines within the scope of the local environmental natural capacity, so as to safeguard the city's sustainability and livability.

Raise Public Awareness of Ecological Boundary Lines The ecological boundary lines of urban areas are broken in the process of urbanization and industrialization. One reason is that urban societies focus on economic interests and lack awareness of ecological boundary lines. When deciding local industrial structure and resource utilization, powerful governments focus on wealth accumulation, economic growth, and fiscal revenue. Environmental debts are not included in government accounting and performance assessment. In their operations, enterprises often focus on the profit maximization pressure or temptation and neglect their environmental protection responsibilities. Some enterprises have their own water wells to extract underground water above the limit set by the government. To save cost, they do not invest in water conservation technologies and equipment. They focus on short-term cost and interest and neglect the total quantity and speed boundary lines. Consumers are both victims of broken ecological boundary lines and accomplices of the problem. Land resource is in shortage, while many consumers take real estates as investment goods and speculate on them. This leads to increase in housing prices and waste of land resources. Each car user has his or her share of responsibility for traffic congestion, air pollution, and greenhouse gas emissions. The governments, enterprises, and consumers all think that environmental boundary lines are restrictions for all but themselves. They consider themselves victims of others' actions, blame others for the environmental problems, and expect others to take action. If all social members lack awareness of the ecological boundary lines and refuse to take action, then it is impossible to safeguard the ecological security.

Strictly Safeguard the Ecological Boundary Lines First of all, it is necessary to clarify the issue through legislation and strengthen law enforcement. Excessive underground water extraction, occupation of green land for other uses, and unconstrained pollution emissions happen because the importance of ecological boundary lines are not specified through legislation. Laws are rigid rules for human behaviors. It should be stipulated in law that if ecological boundary lines are broken, it breaks the law and the concerned people will be punished accordingly. Boundary line of water resource in Beijing has been continuously broken and the city's water resource deficit is continuously growing. But the municipality of Beijing is seeking water diversion from other places, rather than improving water resource protection and restricting water utilization through legislation. With the worsening of smog, the municipality mainly relies on administrative measures, instead of legal measures, to reduce smog. Second, city profile and structure of industries should be considered in urban planning. When planning their spatial and industry development, super large cities mainly make the decision based on administration and economic costs and benefits. They do not attach enough importance to environmental capacity boundary lines. For example, some cities divide their urban areas into educational zones, industrial zones, cultural zones, medical and health zones, residential zones, business zones, and so on. Between zones, there are clear boundaries; such zone division makes it easy for government administration. However, it also leads to the separation of living places and working places and different allocations of different functional zones. Furthermore, super large cities also use their power in resource allocation to monopolize the best resources, which attract the population inflow and concentration in these cities, causing continual outward expansion of these cities. To solve the problems, in super large cities, urban planning should consider mixing living places and working places, and overlapping different functional zones, giving up some secondary functions. If super large cities continue using their power in public resource allocation and developing all functions, instead of giving up some secondary functions, the efforts to control the population and size of these cities will be in vain. Thirdly, the government should use powerful economic instruments to regulate consumption demand and safeguard ecological security. The housing prices are very high in super large cities. If a property tax is levied based on the size and value of the real estate properties, the situation of housing speculation and large amount of housing resources unused in cities will be effectively changed. Tiered pricing for water resources, electricity, and oil could effectively stop resource wasting and pollutant emissions. Actually, the forming and enhancing of environmental protection awareness needs effective implementation of laws and policies. Following the law of nature and complying ecological boundary lines can guarantee the realization of proper environmental governance and protection of human habitations.

9.3 Ecological Compensation Mechanism

Under the institutional framework of industrial civilization, market prices to some extent reflect the compensated use of natural resources which have market demand and supply. But in most cases, this market price level only reflects value of labor instead of resource scarcity and regeneration/substitute cost. Paid use of resource requires that human life and production activities shall promote sustainable use of natural resources and normal functioning of ecosystems. It requires people to pay for enjoying ecological service and consuming natural resources according to the resource value and scarcity.

9.3.1 Ecological Asset with Natural Increment

The ecological compensation mentioned here is the property using fees paid for natural resources and ecological asset owners or royalties and ecosystem maintenance fees for ecological service provision. Therefore, ecological compensation, in reality, is a kind of compensation to ecological asset owners, including asset income sharing and compensation for the opportunity cost loss. In principle, ecological asset owners shall comply with laws and regulations related to natural resource conservation. They have no rights to destroy natural resources and should not be compensated for causing no damage to ecological environment. On the other hand, ecological asset owners' sustainable development will be endangered if they damage ecological service functions. But under the socialist system in China, the government does not allow for private ownership of land and other ecological assets; it only accepts private rights of use. Furthermore, socialism seeks common prosperity. It should also protect vulnerable ecosystems and the development rights of the legal users and beneficiaries of ecological assets who provide ecological services. Therefore, ecological compensation, in essence, is the market compensation for the opportunity cost loss of people who have ecological asset using rights, in order to safeguard the ecological service functions of ecological assets. Under the socialist legal framework of natural resource ownership, ecological compensation system innovation needs to construct mechanisms and systems for transfer payment to key ecological function zones, inter-regional payment for ecological services, and ecological compensation based on the market demand and supply of ecological services.

In terms of the natural conservation zones, water conservation zones, wetland water systems, and forest ecosystem with significant ecological functions, the users and scope of their ecological functions are obviously not restricted to local area. But residents in these regions have for many generations been relying on the output functions of local ecological systems. In order to protect and exert the ecological service functions in these zones, local residents have to forsake the development opportunities of polluting and destructive industrialization and urbanization which could give them high short-term economic returns. In reality, in order to protect ecological environment, the government restricts the industrial development and use of key ecological function areas. Such restrictions protect ecological barrier and sustain ecological service, but they also impede the free trading of ecological resource as a production factor. Consequently, the economic returns in ecological function zones are lower than those in eastern and developed areas. The income of ecological function areas is low because of the insufficient compensation for their contributions to state ecological environmental security. In this sense, the central government's transfer payment to poor ecological function zones is neither "poverty alleviation" nor "charity." It is a payment for ecological services and ecological compensation.

The government takes the main responsibility of ecological environmental protection; but this doesn't mean that the government should cover most of the expenses. According to the principle of ecological compensation, "the developers should protect ecosystems, the destroyers should restore ecosystems, the beneficiaries should compensate for the use, and the polluters should pay for the pollution." Therefore, who shall cover the expenses is a problem about responsibility sharing among stakeholders. The essential connotation of "ecological compensation" is that beneficiaries of ecological service function should pay suppliers for the ecological service functions. So the beneficiaries can be government agencies, individuals, enterprises, or a region. It requires the government to establish relevant systems, laws, and regulations to clarify the responsibility, power, interests of different parties, and their interrelations and promote marketization of ecological compensation and public participation. The government should set up mechanisms and enact policies to require the developed regions which receive ecological service and protection from ecological barriers to make ecological compensations to the regions which provide these services and protection. For example, the interests of regions in a watershed are naturally interconnected. If the surface water and underground water flowing from upstream are not polluted, the downstream areas have some moral and even substantial obligations of giving ecological compensation to upstream areas. Some ecological service functions can be compensated through market, for example, the market price of tourism, ecological function products such as natural spring water, wild foods, and so on. For example, the national government can help increase the fiscal income of local governments and job opportunities for local residents through supporting these regions use their advantage of good ecological environment to develop such industries as ecological agriculture, ecological forestry, ecological tourism, and renewable energy development. It can offer these regions some preferential policies, for example, more bank loans, fiscal interest subsidies for loans, investment subsidies, and tax reduction, so as to promote the continuous development and growth of characteristic industries in these regions within the carrying capacity of local environmental resources.

China should establish various compensation methods and explore diversified compensation methods such as monetary compensation, compensation in kind, policy compensation, and talent compensation. The government should encourage multiple compensation methods on the basis of transfer monetary payment. This could be counterpart cooperation between the ecological service beneficiary regions and the ecosystem protection areas, targeted construction support for hardware and software infrastructure facilities, industry transfer and industry value chain extension compensation, development in different regions, and joint industrial park development. A mechanism for long-term inter-province, inter-region, and inter-watershed ecological compensation should be established. The fiscal-dominated ecological compensation should be combined with such market mechanisms as carbon sink, emission trading, water rights trading, deposit and refunding system, and eco-labeling. The financial support should be combined with talent cultivation, employment training, technical assistance, and industry support. These combinations are aimed at maximizing the effectiveness of ecological compensation.

The maintenance and exertion of ecological functions is an autonomous process of ecosystems. Even ecological function zones with fragile environment have self-restoration and increment capabilities. Reducing and even relocating all productive population in the relevant functional areas are the most effective ecological compensation. Industrialization in East China has attracted a lot of labor from the ecological functional area in the past few decades. However, due to the rigid household registration system and local protectionism, the transferred agricultural population from ecologically vulnerable areas cannot get full access to public social services and social benefits in the cities they work and live. For example, eight million people with hukou registered in Guizhou Province in Southwest China are working and living in East China. Although the primary motivation of these people's migration is not to reduce ecosystem destructions, their migration does reduce the ecological pressure in Guizhou. If these people ultimately have to return to Guizhou, their hukou registered place, because they are not accepted as equal urban residents in East China, monetary compensation cannot fundamentally improve the ecological environment in Guizhou. Eliminating the barriers for the migrant farmer workers to move their hukou registration from their hometown to the places they work and live can be a key approach for ecological compensation. Guizhou provides ecological service and transfers such ecological assets as clean air and water, stores carbon sink, and protects biodiversity: provinces in East China absorbing rural migrant workers also use and benefit from these ecological services and products. Therefore, from the perspective of inter-regional ecological and economic cooperation and in view of targeted ecological compensation among different provinces and areas in a watershed, the coastal provinces have both duties and obligations to give priority to migrant workers from Guizhou when accepting hukou registration of people from other parts of the country. This can avoid the eight million economic migrants returning to their hometown in ecologically vulnerable areas, which can not only avoid ecological damage in Guizhou but also avoid payment of poverty alleviation fund.

The low industrialization and urbanization level in ecologically vulnerable area in West China is partially due to the limited developable spaces in these areas. For example, the landscape and topography of Yunnan, Guizhou, Sichuan, and west of Hunan and Hubei provinces are mainly high mountains and steep slopes. These areas are not competitive in industry development and their city development is restricted by land conditions. For these areas, planning and implementing ecological migration for economic development and establishing of "exclave industrial parks" in other regions for industrial and urban development are better options than spontaneous migration and local citizenization of farmer workers. To compensate the ecological service providing regions for their lost development opportunities due to environmental protection, they should be allowed to establish "exclave industrial parks" in downstream areas with big industry clusters and relocate enterprises and labors to "the exclave industrial parks" for economic development. The origin and host areas should share the taxes and development dividends of "the exclave industrial parks." Meanwhile, in the water source areas or key ecological functional areas where industrial enterprises and labors move out, the governments should focus on developing industries with ecological restoration effects, such as the tertiary industry, tourism, green agriculture, and characteristic mountain industry to create new engines for ecological economy development. In this way, the "one-way blood transfusion" can be changed into "double blood making." It can also realize the dynamic combination of establishing ecological compensation system and supporting important eco-functional area development, achieve the all- win of eco-functional area environmental protection, industrial agglomeration, energy conservation, emission reduction, and green transformation of economic structure.

Investment subsidy for industrial upgrade and extension of ecological industrial chain are also effective ecological compensation methods. In ecologically vulnerable areas, there are many national geological parks, forest parks, scenic areas, and tourist destinations, where the ecological services and products have market demand and can be sold in the market. These areas are short of funding, technology, hardware, software, and other infrastructures for industrial transformation and upgrade. The national government should provide investment subsidies for the industrial upgrade in these areas and help them construct roads, hotels, and other infrastructures, as well as such social service facilities as schools, hospitals, and cultural facilities, so that the ecological compensation can be converted into these areas' environmental protection and development capabilities. The stakeholders of regional ecological compensation should undertake industrial chain complementation and extension through signing long-term purchase contract and setting up cooperative factories and joint ventures. For example, in Guizhou Province, Bijie County is in the upstream area of Renhuai County, where the Moutai Distillery is located. In order to make sure that the water quality in Renhuai County can meet the requirements of Moutai Distillery for making spirit, Bijie has to forgo development opportunities that may pollute its water. In addition to monetary compensation to Bijie County, through signing long-term purchase contract, Moutai Distillery can guide the upstream areas to undertake the industrial chain of distillation, such as growing and supplying sorghum, main raw material for its spirits production, producing packages for its spirits, so as to form an ecological community with the upstream areas.

Market instrument can also be used as an ecological compensation approach. The cost of domestic sewage treatment is over one yuan per tonne. The annual operation cost of a wastewater treatment plant with a daily treatment capacity of 30,000 tonnes is more than 100 million yuan. If the ecological purification functions of wetland are utilized to treat the household sewage, it can not only save large amount of sewage treatment cost but also support the growth of many green plants, which can be used as carbon sinks and alternative energy resources. Enterprises with carbon emission exceeding the national standards can purchase carbon sinks or invest in tree planting to conduct ecological compensation.

9.3.2 Nonrenewable Resources

Though the utilization of nonrenewable resources is not a problem of sustainability, ecological compensation is also needed for three purposes: restoration and reconstruction of surface ecosystems; research, development, and use of alternative resources; and value of resource owner's equity. The Chinese laws stipulate that the ownership to all underground minerals belongs to the state; thus, the owner's equity can be realized by the government through coercive measures. However, the exploitation of underground minerals inevitably involves the development and utilization of land resources and compensation to the land users for their interest loss. As in China mineral resources are mainly explored by the state, corresponding ecological compensation is often ignored and large amount of ecological debts have been accumulated. This problem is especially prominent in the cities where mineral resources are exhausted. Due to ecological environmental crisis and singular economic structure, these cities face many barriers in economic restructure and ecosystem restoration. Obviously, the government needs to set up long-term compensation mechanism for the development and utilization of nonrenewable resources.

Resources exploitation has made huge contribution to the country's development. However, in the cities where resource exploitation takes place, enormous social and ecological debts have been accumulated. With the decrease of resources reserves and the increase of exploitation costs, the marginal profit of resource exploitation will continuously decrease. The resource-based cities with singular industrial structure face the problem of resource depletion, loss of development engine, serious ecological damages, lack of restoration capacity, slowing down of economic development speed, social instability, and other difficulties. Take Shizuishan in Ningxia Hui Autonomous Region as an example. Decades of coal mining and processing have caused seven major environmental destruction problems to the city, including large quantity of solid wastes, emission of harmful gases, ground collapse of goafs, water system damage, water body pollution, vegetation deterioration and biodiversity loss, land resource occupation with no returns, as well as soil erosion. Since these problems have not been taken seriously and solved synchronously in the course of resource development, their continuous accumulation leads to ecological crisis in the cities where mineral resources are exhausted. In the economic and social aspect, these cities face the problems of singular industrial structure, inappropriate urban spatial layout, and imperfect social security system. Because of these problems created during their development process, these cities, during their declining stage, face the difficulties of lacking strong alternative industry, shrinking job opportunities, absence of development driving force, and increasing discontent among local people. During the formation and growth stages of resource-exhausted cities, these problems were often masked by rapid economic growth and high urbanization level; after they enter the declining stage, these hidden problems suddenly show up, which severely affect the social stability, economic development, and ecological safety in these cities.

Actually, the "location advantages" of resource development, to a large extent, means local ecological and social cost. The costs of resource development stay behind, while the benefits shift to other places and regions. In Ningxia, the areas with large coal reserves are sparsely populated and the compensation payment it receives for ground and space occupation due to coal mining is low. Under such understanding about "location advantages," ecological damage and the cost of society development are considered as revenue and become profit of investors and national tax revenue. Meanwhile, the costs continuously accumulate in resource-based cities, causing and deteriorating the social and ecological crisis in these cities when their coal reserves are exhausted. Ningxia generates electricity from coal and sells it to other Chinese provinces. The benchmark price of coal-fired electricity from Ningxia is much lower than the electricity importing provinces, which is regarded as the "location advantage" of Ningxia. Take thermal power transmitted in direct current from the Ningdong City in Ningxia to Shandong as an example. The electricity price is fixed as 0.3213 yuan/kWh, while the benchmark price of desulfurizated and denitrated coal-fired electricity supply for local use in Ningxia is only 0.2841 yuan/

kWh. The final electricity tariff for grid access in Shandong is 0.4356 yuan/kWh; while the tariff for desulfurizated and denitrated coal-fired electricity generated in Shandong is 0.4432 yuan/kWh. Since the ±660 kV direct current transmission line from Ningdong to Shandong was put into use in 2011, the China State Grid Company had transmitted 100.238 TWh of electricity from Ningxia to Shandong by the end of June 2014, accounting for 9 % of the electricity consumption of the whole Shandong Province. The electricity purchase helped Shandong avoid 78.7 million tonnes of CO_2, 220,000 tonnes of SO_2, and 193,000 tonnes of NOx. Taking into account the electricity transmission and transformation costs, Ningxia's electricity generation cost is around 0.1 yuan per kWh cheaper than that in Shandong. The benefits from the "comparative advantage" of Ningxia's low electricity generation costs for the 100.238 TWh are over one billion yuan. They are profits shared by the electricity generation enterprise, the grid company, and the end users in Shandong. Meanwhile, the pollutants are left in Ningxia. The responsibilities of social and economic transformation and ecological restoration after resource exhaustion are also left to Ningxia. Obviously, a large proportion of such transfer earnings is the cost of social and economic transformation and ecosystem maintenance and restoration of resource-exhausted areas and shall be retained in Ningxia.

Compensation relying on fiscal transfer payment is "afterward old debt repayment," and compensation is made by the government, not the beneficiaries of the resource depleting activities. It was for the first time clearly stipulated in 2001 in the national "10th Five-Year Plan" that development of subsequent industries and alternative industries shall be promoted in resource exhausted cities. In order to promote transformation of resource-exhausted cities, the central government makes enormous fiscal transfer payment to these cities for industry transformation, ecological restoration, and infrastructure construction. In 2013, the total transfer payment from the central government to resource-exhausted cities reached 16.8 billion yuan, 5 % more than that in the previous year. Such fiscal transfer payment to resource-exhausted cities is essentially "old debt repayment," namely, after-event remedial measures to repay the debt of local ecological and social damages. As one of the first-batch resource-exhausted cities, Shizuishan got more than two billion yuan of fiscal transfer payment from the central government since 2007; that is to say, on average, it received more than 300 million yuan of subsidy per year. Most of the fund has been used for retrofitting the urban slums and coal cinder hills, management of goafs and collapsed area, relocation of the mining areas, and infrastructure construction for the development zone in Shizuishan. The relocation project alone has cost about 600 million yuan over the years. This shows that most of transformation subsidies are used for "old debt repayment," the subsidies are unable to effectively support the local industry restructuring and subsequent industry development.

In such a compensation approach of vertical fiscal transfer payment, the compensation comes from the state. The resource enterprises that cause huge environmental damages to resource-based cities and resource purchasing areas that obtain enormous profits from resources exploitation are exempted from undertaking corresponding compensation responsibilities. Thus, the fundamental principles of "resource developers protect resources, beneficiaries compensate the costs, and polluters restore the environment" are not actually enforced. The restoration costs to eliminate ecological damage are high and many environmental damages are irreversible. Therefore, post-damage compensation for resource-exhausted cities is just a temporary solution, not a permanent cure.

The existing ecological compensation system in China lacks a mechanism for long-term effect. Government departments at all levels have tried various approaches for establishing an ecological compensation mechanism for resource development. Take coal resources for an example. The coal-producing provinces raise contributions to a coal sustainable development fund and require the mining enterprises to pay environmental restoration deposit according to the *Mineral Resources Law*. The two funds have different withdrawal and application procedures and regulations. But both funds have some weaknesses in raising, application, and management. Firstly, in the absence of relevant laws and regulations, both funds encounter multiple obstacles in the process of raising and management; to prevent misuse, their monitoring and management supervision require enormous human and material resources. Except the coal sustainable development fund of Shanxi Province, which is approved by the National Development and Reform Commission, various funds of other provinces are approved by provincial, municipal, or county governments, and their rules and regulations are also set by local governments. Frequent changes in the contribution rates and lack of relevant national regulations and supervision make the collection of these funds unstable and lack a mechanism with long-term effects. Secondly, collection of coal fund has to some extent increased the fiscal revenue of local government, but there is no effective monitoring and supervision on the direction and place of their uses. Thirdly, for mine environmental restoration and management deposit, use of this fund is limited to ecological restoration and environmental management of mining areas where the enterprise is located. It does not cover the ecological damage beyond the boundary of the mining area caused by mineral resource exploitation and inter-regional environmental pollution.

The "comparative benefit" of resource utilization shall be used to cover the cost of social transformation and ecological damage of resource-based cities to synchronize the damage and compensation and to form long-effect mechanism. The current approach of afterward compensation for resource development is passive transformation after the occurrence of resource exhaustion, ecological and environmental damages, and social crisis. The transformation practices of resource-exhausted cities in recent years indicate that the cost paid through the afterward compensation approach is huge, while the effects are often unsatisfactory. For resource-based cities that have not entered their declining period, such as the Ningdong Energy and Chemical Industry Base, in the process of their development and construction, part of the "competitive benefit" due to the "location advantages" shall be earmarked and reserved for long-term ecological compensation and social compensation. The fund shall be used for industry diversification, ecological protection, and social security so that the resource-based cities that rely on a single resource industry can be changed into ordinary comprehensive cities.

In this way, the compensation no longer comes from the central government, but the direct beneficiaries of resource utilization. Such compensation is synchronous with environmental damage; it will not cause accumulation of environmental damages and delays in solutions. In addition, such compensation has long-term effect and will not be interrupted. Of course, the compensation revenue shall be spent with proper planning and take into account future needs, instead of focusing on short-term interests. Resource ownership shall be clearly defined to ensure that considerable part of benefit from resource development is used for ecological and social compensation. The *Mineral Resources Law* specifies that mineral resources belong to the state and the State Council exercises the ownership of mineral resources on behalf of the state. However, in practice, state ownership of resources actually falls into the possession of exploitation and utilization enterprises. The confusion of ownership and the rights to utilize and operate resources has caused serious separation of beneficiaries and compensation payers in the process of resource development. Recognizing that compensation is actually to recognize the rights and interests of local residents and ecosystems in mining areas. Defining such recognized rights and interests as an integrated part of resource ownership and establishing and improving corresponding laws, regulations, and policy system are keys to define the main payers of resource exploitation compensation mechanism. This is also a fundamental precondition for establishing long-effect and stable compensation mechanism.

9.4 Ecological Governance

China's current environmental and ecological protection administration system is refined top-down controlling administration based on industrial division under the industrial civilization. The central government and local government at different levels set corresponding functional departments, and the administration is carried out through policies, directives, standards, and other means. Although ecological damage is serious in China owing to the country's huge population pressure and shortage of arable land resources, there is no governing authority focusing on ecological protection. With deterioration of environmental pollution and increasing conflicts between environmental protection and economic growth, the environmental protection agency in China has been continuously enhanced and upgraded. In 1972, the Chinese government sent a delegation to attend the First United Nations Conference on Human Environment. It then established an Environmental Protection Office under the National Construction Commission in 1973. Fifteen years later, in 1988, it was separated from Ministry of Construction and the State Environmental Protection Bureau was established and put under direct administration by the State Council. Another 10 years later, the State Environmental Protection Bureau was upgraded to the State Environmental Protection Administration in 1998. In 2008, after 35 years of development, the State Environmental Protection Administration was further upgraded into the Ministry of Environmental Protection and became part of the State Council. The Ministry consists of departments and bureaus on such issues as policies and regulations, scientific and technological standards, pollutant cap control, environmental impact assessment, environmental monitoring, pollution prevention and control, natural ecological protection, nuclear safety administration, and environmental monitoring. These departments are also subject to top-down administration inside the ministry. At the same time, institutions and functions related to resources, environment, and ecology administration are also established or strengthened inside traditional macro administration authorities and natural resources utilization administration authorities, such as the Resources and Environment Department and Department for Addressing Climate Change under the National Development and Reform Commission, the Ministry of Water Resources, the Ministry of Agriculture, the Ministry of Industry and Information Technology, the State Forestry Administration, the National Energy Administration, the State Oceanic Administration, and China Meteorological Administration. Environmental protection departments have been established in local governments at provincial level and environmental protection bureaus in governments at prefecture and county levels. Environmental protection bureaus have been set in cities (districts, counties) of all provinces and municipalities directly under the central government. In their environmental protection administrative functions, the Chinese environmental protection agencies follow a vertical administration hierarchy of Environmental Protection Ministry – Environmental Protection Departments – Environmental Protection Bureaus. The human resources, budget, and funding of local environmental protection agencies are administrated based on jurisdictions. In other words, the local governments are in charge of the size of personnel force, appointment and removal of environmental protection officials, and allocation of funds for their environmental protection departments and bureaus.

As for the environmental administration laws and regulations, systems, instruments, approaches, and mechanism, the National People's Congress, environmental protection authorities, and relevant government authorities have formulated, amended, and improved a series of laws, regulations, and policies. The examples include *Environmental Protection Law, Law on Prevention and Control of Water (Air, Noise, Solid Waste) Pollution, Energy Conservation Law, Law on Cleaner Production Promotion, Regulations on the Administration of the Collection and Use of Pollutant Discharge Fees, Provisional Administrative Measures of Water Pollutant Discharge Permit, Measures for Environmental Protection Supervision and Administration of Sewage Treatment Facilities*, and *Regulations on the Reporting and Registration of Pollutants Emission.* There are other laws and regulations on the emission and discharge standards for various types of pollutants. On the basis of the laws, China has established the corresponding environmental administration systems, such as administrative permit, wastewater discharge permit, pollution discharge fees and their collection, environmental impact assessment, environmental monitoring, "three-simultaneousness," "treatment within limited time," "pollution discharge application, registration and reporting system," "total amount control on pollutants," and "comprehensive improvement of urban environment quantitative assessment." A long list of administrative instruments and tools are used, including environmental monitoring, environmental remote sensing, total amount control of pollutant discharge, monitoring of key areas (industries, emission enterprises), pollution prevention and control, and environmental law enforcement. The key targets of environmental and ecological protection are included as blinding targets in the national Five-Year Plan for economic and social development. Special pollution administration actions and other governance methods are implemented for specific serious environmental problems. The emission trading systems have been set up in some regions. The government sets benchmark prices for the trading of pollutant emission rights. The government also sets up a comprehensive system to guarantee the smooth implementation of daily works, including regularly publishing environmental communiques, environmental information disclosure, public awareness raising and education, international cooperation, receiving and reporting complaints, and other daily environmental administration work. The elements mentioned above are the basic governance framework of current ecological and environmental protection in China.

In general, as far as ecological environmental protection and administration system is concerned, China has formed a relatively complete organization system and laws and regulations system in line with *industrial civilization*. Ecological environmental protection targets have become binding targets in the national Five-Year Plan for economic and social development and important contents in the performance assessment of government officials. However, this system framework can't meet the need of ecological governance.

Firstly, the relevant institutional setup and function division of ecological environmental protection are scattered. Except specialized environmental protection agencies, relevant ecological environmental protection departments and divisions are also set up in the government authorities on comprehensive administration and other line ministries and local government authorities, which have caused replication of administrative organizations and responsibility overlapping among different government agencies. Relevant functions of ecological environmental protection functions, including pollution control, ecological conservation, climate change, water environment, traffic environmental protection, agricultural and rural environmental protection, protection of forestry and wildlife, marine environmental protection, and meteorological environment are scattered in different government departments and organizations.

Secondly, ecological environmental protection laws, policies, and agencies are in relatively weak positions. First of all, compared with economic growth, ecological environmental protection is in weak position. Governments at all levels and in all places give priority to ensure the realization of economic growth targets, and the realization of ecological environmental protection targets are often taken as secondary priority. Second, compared with other functional government departments, ecological environmental protection agencies are in weak position. Because greater importance is attached to economic growth than to environmental protection, therefore among the component departments and divisions of local government at all levels, environmental protection agencies are less powerful and in a weaker position compared with the departments and divisions responsible for promoting economic growth and controlling economic resources. It is difficult for the environmental protection agencies to coordinate with these more powerful departments and divisions. The third is that among government administration functions, the enforcement of environmental protection laws and regulations is relatively weak. For example, to speed up local economic development, some local governments permit some large projects to start construction before they complete and pass the environmental impact assessment. The relevant laws and regulations of ecological environmental protection are totally ignored, leading to destruction of ecological environment and waste of economic resources.

Lastly, ecological environmental protection administration is relatively lax in general. First, the enforcement of environmental standards is lenient. Generally, China's environmental standards are less strict than those of European countries and the United States, which weaken the restraint and regulatory function of environmental standards. Moreover, due to pressure of market competition, cost of environmental protection, cost of environmental law enforcement, and other factors, environmental standard implementation often falls short of the standards' requirements, and the binding effects of environmental standards are faced with high resistance and pressure. The second factor is insufficient environmental monitoring and supervision. Due to the manpower, material resources and financial resource restrictions of grassroots environmental protection agencies and the large number of diversified pollutant emission entities, emission sources and pollutants, and vast areas of environmental monitoring and supervision, environmental protection agencies also find it difficult to effectively monitor and supervise the emissions of key enterprises and key sources, comprehensive supervision is even more beyond their capacity. The third factor is lenient environmental law enforcement. In their environmental protection administration, some local environmental protection agencies are not strict in environmental law enforcement, sometimes turn a blind eye to the illegal pollutant emission of enterprises and regard this as an excuse for increasing the collection of pollutant discharge fees and penalties. In some occasions, due to intervention from local governments, environmental protection agencies are unable to strictly enforce laws and regulations on large enterprises which key to local economic growth and with high contribution to local tax revenue.

The fundamental element of the institutional innovation toward ecological civilization is the change of power mechanism. The efficiency of industrial civilization derives from the top-down centralization and authority. Ecological civilization requests the change of governance structure, from authoritative ruling, governance, and administration to the participatory governance or management by stakeholders. According to the definition by the UN Commission on Global Governance, governance is the summation of multiple approaches of individuals and public or private institutions in operating and managing an affair. It is the continuous process to reconcile mutual conflicts or different interests and take joint actions. It includes official institutions and rules and regulations that have the rights to force people to obey all kinds of informal arrangements. All of these are authorized with power by the people and institutions through agreement or confirmation on conformity of their interests.[3] Therefore, unlike ruling and control, governance is a kind of activity supported by the common goal and the government may not be the coordinator and implementer of these management activities and their realization does not necessarily rely on the mandatory force of state. Therefore, the governance and ruling are fundamentally different. The authority of ruling and control mainly comes from government. Although governance also needs authority, the authority is not monopolized by the government. Governance is the cooperation between the state apparatus and civil society, governmental and nongovernmental organizations, public institutions and private organizations, as well as compulsion and freewill. Government authority administration is top-down and single-dimensional management of social affairs through policy making, directive guidance, and standard regulation. In contrast, governance is the interactive process of multiple stakeholder participation. Governmental agencies, nongovernmental organizations, and various private entities jointly manage public affairs through cooperation, negotiation, and partnership and by means of common goals. The rights of managing social affairs are exerted by multiple actors and through multiple dimensions, including bottom-up and top-down approaches. The role of social forces is strengthened in governance, and various social forces can have bottom-up influences on public authorities through all kinds of effective approaches.

The theory of self-organized governance put forward by Eleanor Ostrom further weakens the authority of government. Self-organized governance is that a group of interdependent agents organize themselves for self-governance to realize their lasting common interests under circumstances that all members are faced with free riding and avoidance of responsibility or other opportunistic behavior. When Ostrom discussed this problem, she classified the internal variables influencing personal choice strategies under complex and uncertain environment into four factors: expected benefit, expected cost, internal norms, and discount rate.

9.5 Ecological Legislation

China enters the primary stage of socialism from a semi-colonial and semi-feudal society and a new democratic society. In the process, Chinese traditional civilization has been influenced and changed by Western industrial civilization. Changing nature, technological innovation, market competition, economic globalization, and other value elements of industrial civilization as well as its corresponding systems and mechanisms drive the industrialization and urbanization process in China. The Chinese economy has achieved enormous growth and development; meanwhile the country's environmental challenges are becoming increasingly severe.

Even though there are many laws and regulations on ecological environmental protection, ecological environmental construction face the problems of relevant

[3] Yu Keping. (2000). Governance and good governance. Social Sciences Academic Press, pp. 270–271.

laws that are "fragmentizing" and even conflict with and cancel out each other. For example, relevant provisions on resource utilization of solid waste are included in *Law on Cleaner Production Promotion, Law on Circular Economy* and *Environmental Protection Law* as well as some other laws and regulations. The laws and regulations on pollution control and energy conservation are mutually independent, leading to the ignorance of energy saving in pollution control administration and the weakening of environmental protection in energy-saving administration. Many wastewater treatment plants and desulfurization facilities are left idle and unused after construction and installation. Apart from economic interest considerations, different requirements in different laws and regulations are another important reason.

Amendments to laws and regulations are delayed and unable to meet requirements of constantly deepening ecological civilization construction. For example, under strong impetus of public opinions, the concentration of atmospheric particulate PM2.5 was included in the environmental monitoring system in 2011, but control target and measures of PM2.5 seriously lag behind. For another example, the carbon sink function of forest is not reflected in existing *Forest Law*, and there are no regulations to encourage and support carbon sink creation and maintenance. As "instruments for macro adjustment and control," the laws and regulations on ecological civilization construction are not absolute rules with rigid constraint. They are often flexible in enforcement and leave too large discretion room; therefore the enforcement of these laws and regulations is often highly arbitrary. Moreover, the strength of penalties on illegal enterprises and law enforcement efforts are insufficient; sometimes the law enforcers themselves even violate and bend the laws. This weakens the authority of the laws and regulations and the actual effects of their enforcement.

The relevant provisions on ecological civilization construction in laws and regulations are mainly principles and lack detailed operational guidance. To be implementable, relevant provisions need to be refined and supplemented with detailed rules and regulations, ordinances, and policies. But these detailed rules and regulations as well as policies are often temporary and ignore long-term effects. The policies constantly change and are discontinuous; investors and manufacturing enterprises do not know what to do and cannot make long-term business strategies and decisions in accordance with the policies. Taking *Renewable Energy Law* enacted in 2005 as an example, the full text of this law is less than 4000 Chinese characters and contains few detailed rules on implementation. In contrast, the draft *American Electric Power Law* prepared by the American Senate in 2009 contains very detailed rules. It explicitly stipulated that for CO_2 emission trading, the minimum price is 12 dollars (with an inflation adjustment of 3 % increase per year) and the maximum price is 25 dollars (with an inflation adjustment of 5 % increase per year).

The environmental administration systems are fragmented among different ministries, government departments, and between central and local governments. It makes it difficult to hold any specific government agency responsible and accountable. The different governmental authorities compete for benefits, avoid risks, and shed responsibilities. There are interactions and coordination mechanism in the administration of major issues and projects for ecological civilization construction.

However, each government authority has its own interests and considerations. In the process of implementation, specific requirements of ecological civilization construction are often ignored in "power and interest" gaming between different government departments at the same level as well as between central and local institutions. As for administration mechanism, green, circular, and low-carbon developments have been identified as approaches for ecological civilization construction, and corresponding plans and targets have been set for them. However, the mechanisms for target allocation, specification, and performance assessment, monitoring and supervision, compliance, and public participation have not been comprehensively established.

Another issue is intra-generation and intergeneration compensation. As an important incentive mechanism for ecological civilization construction, the ecological compensation mechanism lacks specific legal orientation, legal basis, and market mechanism. In terms of intra-generation compensation, the upper stream areas of a water basin have legal obligations to protect local ecological environment and such protection activities shall not be conditioned to receive compensation. In order to obtain compensation, some areas intentionally damage their local ecological environment. This is contrary to the original legislation intention of ecological civilization construction. And the compensation is not strictly based on "market supply and demand" due to the lack of price elasticity; instead it is interest ruling based on market evaluation and has legal binding effect. If it is "purchase of ecological service," then it is a kind of market contract with price elasticity. Due to public goods property of ecological service, "governmental purchase (transfer of payment from central government or local government)" or collective purchase (government, social groups, or enterprises in lower streams of the river, such as water supply enterprises on behalf of water user groups), has the dual attributes of ecological compensation and ecological service purchase. As the future generations cannot influence the current decision-making, the intergeneration compensation is actually a kind of moral obligations and self-discipline, whose implementation also needs to be based on legal provisions and market mechanism.

In the various laws and regulations on resource conservation, environmental protection, and ecological construction, there are provisions on the security of energy, water, food, environment, and ecosystem. But most of the security of these factors are defined and regulated in a narrow sense as well as the nexus and interdependence among energy, water, food, environment, and ecosystem. For example, ecological security mentioned in the laws and regulations on forestry, biodiversity, and wetland, the important contents of inherent linkages of ecological security with food, energy, and pollution control are either barely mentioned or even completely ignored. The ecological security in the laws and regulations is in a narrow sense. But the ecological security required by ecological civilization construction is in broad sense and it covers important nexus with other factors, such as energy, water recourses, arable land protection, and pollution control. The ecological civilization can be integrated into every aspect and whole processes of economy, politics, culture, and social construction only in this way. The 18th National Congress of the Chinese Communist Party requires enhancement of ecological civilization construction.

To overcome the system, institutional, and mechanism barriers that impede ecological civilization construction, it is necessary to recognize the urgency of the task and deepen, improve, expand, and safeguard the progress and development of ecological civilization construction mechanism.

Firstly, it is necessary to enact overarch and comprehensive laws and regulations to promote ecological civilization. In the first political study of the 18th Political Bureau of the Chinese Communist Party Central Committee, Chinese President Xi Jinping pointed out that "with the deepening of national economic and social development, the position and role of ecological civilization construction become increasingly prominent. The 18th National Congress of the Chinese Communist Party had included ecological civilization construction in the overall framework of socialism with Chinese characteristics and thus highlighted the importance of ecological civilization construction." Currently, the laws and regulations related to ecological civilization construction in China are fragmented and lack operability and their contents cannot meet the requirements of ecological civilization construction. One framework law is urgently needed to implement the basic national strategy of resources conservation and environmental protection, put into action the principles of giving priority to resource conservation and environmental protection, and focus on natural restoration and ensures the realization of green, cyclic, and low-carbon developments.

Secondly, it is necessary to set up a national leading group and an advisory committee for ecological civilization construction. Members of the leading group should cover various aspects of economy, politics, culture, society, and ecological civilization construction. The secretariat of the leading group should be established in neutral institutions without departmental benefit or interest conflicts, so as to ensure that ecological civilization construction is integrated into every aspect and whole process of social and economic development. The advisory committee mainly consists of experts and scholars in relevant fields of ecological civilization construction to provide decision-making advice and scientific support for ecological civilization construction.

Thirdly, China should compile behavior guidelines to guide and regulate ecological civilization construction efforts. Ecological civilization has pervasive meaning and universal value. As a member of international society and a strong component of world economy, China inevitably needs to rely on both national and international resources and markets in its efforts to construct a well-off society. When Chinese enterprises invest and do business abroad, they need to follow the rules of ecological civilization construction wherever they are in the world to guide local sustainable development and contribute to global ecological security. When foreign enterprises invest and operate in China, they need to comply with the Chinese laws and regulations on ecological civilization construction to safeguard the local balanced, harmonious, and sustainable development and become a contributor to the movement of building Beautiful China.

Fourthly, China needs to deepen and implement policy of "green" economy development. The "green" economy policies have such advantages as promoting technology innovation, strengthening market competitiveness, and reducing governance and administrative monitoring cost in the course ecological civilization construction. The "green" economy policies shall be specific, predictable, and long-term and operable and can provide continuous driving force for ecological civilization construction. The Chinese government should speed up the establishment of policy framework and system for ecological civilization construction, improve the fiscal system for ecological civilization construction; explore and establish an independent tax and charge system to promote ecological civilization construction; establish resource right and emission trading system at national, industrial, and regional levels; and promote the greening progress of capital market.

Fifthly, China should establish and improve an ecological compensation mechanism. It should specify the legal principles and basis for intra-generation and intergeneration ecological justice and ensure it in law through implementing ecological compensation mechanisms. Funds should be raised from various sources and by various approaches for ecological compensation funds. It can be raised through fiscal transfer payment from the government, payment by ecological beneficiaries, and payments by ecological service and product users, ecological taxes, and social donations. Ecological compensation involves complex stakeholder relations. To balance the interests of different stakeholders and safeguard social justice, China should explore and establish ecological compensation standard system, the rules on the funding sources of ecological compensation, compensation channels, and compensation methods, as well as relevant enforcement and compliance system.

Sixthly, China should establish an assessment and evaluation system of ecological civilization construction and promote "green" accountability system and living style of ecological civilization. It should set up target system, assessment methods, and compliance mechanisms that reflect requirements of ecological civilization. It should also strengthen environmental supervision and improve the accountability system of ecological environmental protection and compensation system of environmental damages. The residents' awareness of resource saving, environmental protection, ecology, low-carbon, and "green" development shall be raised by multiple channels to form the lifestyle of ecological civilization. The government should create low-carbon living environment and guide healthy lifestyle. It should encourage energy saving and emission reduction among consumers through tax reduction and exemption, providing fiscal subsidies and other measures and realize low-carbon life. It should establish and improve the institution system to promote low-carbon consumption, create a low-carbon social atmosphere, and guide the public toward moderate and low-carbon consumption. It should also carry out demonstration and pilot projects and establish some "green" and low-carbon schools, communities, enterprises, and cities.

Chapter 10
Outlook on the New Era of Ecological Civilization

The green transformation toward an ecological civilization is a process. Driven by the unceasing innovation on technologies, the continuous accumulation of capital, and the increasingly consolidated institutions and mechanisms, the transformation from an agricultural civilization to an industrial civilization was realized in countries over a period of one century to three centuries. This transformation has brought countries into an industrial society of abundant material wealth and major improvement in living quality. On the other hand, however, human living environment and resources are still continuously degrading, and the environment is polluted. The era of ecological civilization is definitely not going back to the agricultural civilization times when productivity was low and people struggled for subsistence. Instead, it is an abundant, quality, and sustainable society of ecological prosperity and a steady-state economy within development boundaries and full of society with high quality and a stable economy with vitality but also within resource and environmental boundaries. With the institutions and norms of ecological civilization guaranteeing the international governance of ecological security, countries can fully absorb industrial civilization's advantages and abandon its inherent disadvantages and enter the new era of ecological civilization.

10.1 Ecological Prosperity

The development of human society is motivated by the search for prosperity. Ecological prosperity is not a simple form of material prosperity but a kind of ecological prosperity, or prosperous ecology, and a kind of prosperity with harmony between human and nature.

Productivity and productive relations are essential to the understanding of human society development; the development of productivity is to obtain more material output, so as to meet the need for human society development. The adjustment of productive relations is to emancipate productive forces and to adjust the distribution

J. Pan, *China's Environmental Governing and Ecological Civilization*,
China Insights, DOI 10.1007/978-3-662-47429-7_10

relations so that more members of the society can enjoy the fruits of development. Whether it is productive forces or productive relations, they are both aimed at speeding up the realization of a prosperous society. Industrial civilization has brought about tremendous material wealth to the human society. As pointed out by Walt Rostow, an American economist, to some extent, a society in the stage of mass material consumption, to some extent, has also realized prosperity in material wealth. However, today we realize that this is not the kind of prosperity we hope for and it is unhealthy, unecological, and unsustainable.

The ecological prosperity that we hope for should be, firstly, in line with the ecological law. The components of the ecosystem have their own natural laws in functioning. Production, consumption, and reproduction should be all proportional to maintain a balance. When excessive consumption exceeds the carrying capacity of ecosystem, it not only damages ecosystem functions but also leads to quick ending of the prosperity. A human being, as a part of the ecosystem and as an ecological individual, also has his or her own eco-balance. Excessive food intake and insufficient exercise will result in imbalances in a person's body functions, thus making the individual plagued by illness. Therefore, the systematic consumption of biological cycle is not "the more the better." This also indicates that an ecological prosperity in material is not infinite and does not necessarily need to be maximized.

Secondly, it must be eco-friendly. It is unwise and unsustainable to gain material prosperity at the cost of ecology. The old saying in China "to drain the pond to get all the fish" is obviously a way for rapid material wealth obtainment. However, if the ecosystem suffers such destruction, the prosperity gained can only be temporary. Hence, material prosperity needs to be eco-friendly, and the production and consumption patterns need to be conducive to the functions and operations of the ecosystem.

Thirdly, material prosperity is needed. Although an ecological prosperity is not featured by material wealth maximization, it should definitely not be deficient in materials. There will be no prosperity if basic social material cannot be guaranteed. Therefore, technological innovation and social incentive mechanisms in the industrial civilization will continue to be effective in ecological prosperity. Poverty is not socialism, neither is it ecological civilization. Abundance in material is a basic requirement for ecological prosperity.

Fourthly, it is common prosperity, not prosperity confined to a single country or nation. Productivity of the ecosystem on the Earth provides the material base for the prosperity of human society. Ecological prosperity does not repel or exclude any country or nation, and it is not a zero-sum game. Of course, common prosperity is not absolute equalitarianism but a diversified prosperity with various forms. The diversification of Earth's ecosystems determines the diversification and differentiation of ecological prosperity. Fifthly, ecological prosperity means not only abundance in material but more importantly in spirit. Material abundance is finite, while spiritual prosperity represents the direction and goals of social progress. Spiritual pursuit is unlimited, while material needs have boundaries; spiritual wealth will continue to increase in value, while material wealth can only depreciate over time. A society without cultural and spiritual prosperity is not an ecologically prosperous society.

Ecological prosperity also means prosperity in ecology. Prospering ecology is the foundation and prerequisite for ecological prosperity. With prosperity in ecology, first of all, there must be a diversified ecosystem. The Earth's landform and landscape, together with light, heat, water, and air, form all kinds of ecosystems with different features. They each have their own functions and provide different products and services. If human being interferes or destroys a natural ecosystem, the unique function and services of the ecosystem will also be lost. Typical examples include the following: during land reclaiming from lakes, wetland ecosystem is lost; when people reclaim land through slashing and burning, they destroy the forest ecosystem, which results in flood, drought, as well as water and soil erosion. Such activities are not only unable to increase the productivity and level of material wealth but also impair the production functions of ecosystems. Another feature of prosperity in ecology is biodiversity. The utility maximization under the industrial civilization has misguided socioeconomic activities, leading to excessive hunting, killing, and harvesting of animals and plants with economic value. This has resulted in the extinction of animal and plant species. In addition, animals and plants without market value have not been able to escape from mankind's predations. As a consequence, biodiversity has dropped markedly.

Regional environmental damage causes regional declines in biodiversity; in contrast climate change affects biodiversity globally. The lack of biodiversity impedes the functioning of ecosystems. Imagine, if ultimately the Earth's ecosystem is left only with mankind, the dominant master of ecosystem, will mankind be able to survive? The third feature of prosperity in ecology is highly efficient ecosystem productive forces and high levels of output. Natural ecosystems can regulate themselves and reproduce themselves. Green plants produce material through photosynthesis; various animals and microorganism consume and transform the energy stored in green plants. It is this accumulation and transformation of energy and material that enables mankind to harvest the output of ecosystems and develop from generation to generation. Without human's interferences and damages to the ecosystem, the ecosystem's material wealth can continuously accumulate, leading to higher productivity. Even fossil energy, which highly depends on industrialized societies, is a product created by material accumulation from natural ecosystem. It is the diversification of the ecosystem and its species, along with the natural productivity of the ecosystems that constitutes prosperity in ecology and becomes the material base of ecological prosperity.

Ecological prosperity and a diversified eco-environment should be interdependent, interactive, and in harmony and reaches a prosperity that makes man an integral part of nature. Only this state can be considered as ecological prosperity under an ecological civilization. Ecological prosperity complies with nature, makes use of productivity of the ecosystem, but at the same time protects and maintains ecodiversity. It has a high productivity, which also ensures the ecological material prosperity. The production and consumption of culture, which is the source of spiritual product, also derive from prosperity in ecology. Poems, paintings, and even music are all influenced by the diversity and complexity of nature. Dinosaurs had been extinct long before mankind appeared; therefore, they are rarely seen in spiritual and cultural products. However, other living creatures such as lions, tigers, elephants, birds, as well as flowers seem to be everywhere in classic artworks; some are even worshiped as totems. The extinction of species results in a loss of biodiversity; as a result, mankind not only suffers a loss in material wealth but also has a hard time in maintaining the production and consumption of spiritual wealth. A prosperous ecology is also conducive to prosperity in ecology. With advances in technology, human being could enhance natural productivity or use scientific cognition to protect natural productivity. These are all conducive to protecting eco-diversity and promoting the productivity of ecosystem. Human societies consciously adopt fishing moratoriums, land follows, crop rotations, and nomadism, in order to allow nature's ecosystem to rehabilitate and restore its productivity. Protecting endangered species and biodiversity enhances the productivity and service level of the ecosystem. A prosperity that unites mankind and nature signifies harmony. Prosperity in ecology is an indispensable basic element of ecological prosperity.

10.2 Steady-State Economy

During times of agricultural civilization, human revered nature and complied with nature in economic development and wealth accumulation. However, one natural disaster would force economic development back to the starting point and the process repeated itself. With the advent of industrial civilization, mankind's attitudes to nature changed and started to transform and conquer nature. Economy continued to develop, and wealth kept accumulating. However, periodic economic crisis, which included the stages of recession, recovery, and growth, became a regular phenomenon and persistent curse. The regular damages on and reverses in social and economic development during agricultural civilization were mainly because of natural factors; while those in industrial civilization are mainly because of human factors. The hypothesis of boundless industrial civilization has brought false expectations of infinite development. If growth is obtained only for the sake of growth, after reaching a certain point, then after the economic development reaches a point, the economic size has reached its boundary and the economic growth and decrease will occur alternatively and the efforts to further grow the economy will be in vain. The ecological civilization, after industrial civilization, seeks ecological prosperity instead of infinite material wealth growth. The ecological civilization is not an economy with infinite size extension. Instead, it will be a steady-state economy with constant improvement in quality, but the material wealth will not and does not need to continuously grow.

Economic growth is material. The consumption and accumulation of material goods is confined by the limitations of the Earth. Therefore, in terms of physical space requirements, economic growth cannot be limitless. In the late 1920s, when the US economy experienced the great depression, Keynesian economists diagnosed that the cause was insufficient total demand. Therefore, their prescription was the government should intervene with the public policy instrument and use fiscal account deficit to invest in infrastructure, in order to generate demand and stimulate and guarantee economic growth. However, infrastructure construction within our planet's limited space cannot be infinite. When the physical infrastructure is close to or even exceeds saturation, investment potential in public infrastructure and other fixed assets will be curtailed. During the 2008 global economic crisis, the Chinese government invested four trillion RMB, much of which was used in infrastructure development. For instance, Wuhan City in Hubei Province started the construction of ten subways thanks to the investment stimulus. Whether it is regional or urban infrastructure, this scale of construction could not and should not undergo unlimited extension. Of course, to maintain economic growth, we can introduce some Z-turns: we could take down these facilities before their time comes or to have them upgraded and rebuilt. However, the problem is that although this could bring about economic growth, the total material wealth is not increased. In addition, nonrenewable natural resources such as fossil fuels, steel, iron, and cement are consumed, and the environment is polluted. Overall the losses outweigh the gains. Even this kind of growth can't go on forever. The growth space has already been saturated during the industrial civilization, and it's unlikely that economic growth can be achieved through further space expansion during the ecological civilization.

Population is another source of and driving force for economic growth. During an agricultural civilization, the productivity was low; therefore, large amount of labor was needed to defend against and comply with nature. Back then people had a short life expectancy and high infant mortality rate due to poor medical technologies and living conditions. In order to meet the demand of labor, high birth rate was common, which resulted in fast population growth. During industrial civilization, labor productivity and life expectancy increased sharply. As a result, birth rate dropped and the population growth rate decreased. China was at the initial stage of industrialization in the 1980s, when family planning and birth control policies were enacted, and they were mandatorily enforced in the rural areas. China is still in the mid to late periods of industrialization, and the country's fertility rate and birth rate have become very low. Even the government has relaxed the one-child policy and allows selective families to have two children; early reactions to the policy indicate that many qualified families will choose to have only one child and the birth rate will remain low. In the era of ecological civilization, mankind will have more control of nature, and human population will remain stable or even decline. The population of large predators in nature, such as lions, tigers, and bears, does not increase infinitely despite their top positions in the food chains. This is how the nature functions; therefore, mankind won't experience infinite growth in population either. In fact, Western Europe and Japan, which are in the postindustrialization development stage, have been experiencing some declines in population.

Although the population is not increasing, the living quality has large room for further increase. In fact, the rapid economic growth since China's reform and opening up is mainly boosted by the increase in living quality instead of population growth. In the era of ecological civilization, countries' material wealth has increased significantly, and their material consumption remains at the saturated level. The boosting effects of material consumption to economic growth will be close to zero or even negative.

During ecological civilization, the space of investment in new construction for economic growth is very limited, human population will remain stable or even decline, and material consumption will be close to saturation. What will the economic output look like in this case? The economic theory under industrial civilization uses Cobb Douglas's production function to predict future economic growth: in which Y represents economic system total production, A total factor productivity, K capital input, L labor input, and α a constant.

$$Y = AK^{a}L^{(1-a)}$$

This function indicates that Y is the increasing function of capital and labor, meaning that increases in capital and labor can boost total economic output Y. According to the analysis above, population growth during ecological civilization will be zero or even negative; hence, labor will have limited or no contribution to the economic material output growth. Capital increase can also contribute to economic production and increase material output. The previous analysis shows that under ecological civilization, there will be no market demand for a bigger material output. Since increase in material output is of little value, it is not effective output. Therefore, from the perspective of material production and consumption, economy in the ecological civilization is of steady state, instead of further expansion.

The steady state in material output and consumption does not necessarily mean zero economic growth. To take a closer look at Cobb Douglas's production function, capital can grow indefinitely. Labor or population is constant, and material consumption can reach saturation, but spiritual and cultural consumption, on the contrary, can continue to increase and will not reach saturation. This means that capital can be invested in nonmaterial production to promote to the total output of economic system, which is to say, to produce cultural or spiritual output. This is partially proved by the fact that, at present, material production and consumption is at saturation in developed countries, yet these countries still have a positive growth rate. The material output of these countries has not increased much; the growth mainly comes from the tertiary industry, namely, service outputs.

If the material output and consumption of an economy has zero growth or remain a steady state, then the economy will have stable material demand from the ecosystem; therefore, the quantity of ecological capital will not decrease. During ecological civilization, is there a possibility where the material consumption of human society exceeds the output of the ecosystem and result in ecosystem degradation? The answer to the question is no, or at least it's avoidable. This is because excessive demand for the ecosystem's output mainly happens during the expansive period of industrial civilization. Large scales of infrastructure investment demand and the demand for a better living quality appear simultaneously and result in huge pressure on and damage to the ecosystem. With technology advancement, the ecosystem's output will continue to increase while consumption approaches saturation. Therefore, in ecological civilization, ecological capital will remain stable or increase.

10.3 Transformation Challenges

In the new era of transition toward ecological civilization, mankind has already realized a step forward from industrial civilization. However, as the industrial civilization continually exists, mankind still needs more innovation to eliminate technology bottlenecks, terminate extravagance, squandering and unhealthy consumption habits, speed up the economic growth in developing countries to eliminate poverty traps, and cooperate at global level to overcome governance dilemmas.

Cultivation and farming technologies enabled agricultural civilization to supply primitive civilization. The invention and application of steam engine broke through the limitation of agricultural civilization and prompted a huge leap in society. To accelerate the transformation to ecological civilization, mankind, first of all, faces technical bottlenecks. Electromagnetic and information technology advanced the development of industrial civilization. However, mankind still needs a revolution in energy production and consumption technologies. This is because fossil energy, the driving force of industrial revolution, still supplies the majority of social energy consumption. Fossil energy is not only a nonrenewable resource but also the main factor of environmental pollution, especially air pollution. Fossil energy combustion also emits carbon dioxide, which is a primary challenge for long-term sustainable development and combating climate changes.

Mature economies that are in the later stages of industrial development have effectively reduced their environmental pollution. Their ecological capital is appreciating, and their population and consumption are all on the way to an environment-friendly transformation. However, low-carbon development is still their largest technology problem. In the early 1990s, the necessity for reducing greenhouse gas (GHG) emissions was explicitly put forward, and in 1997, the Kyoto Protocol that limits the absolute GHG emissions of developed countries was agreed. Since then, developed countries have made great effort in emission reduction, but the resulted emission reduction was small.

The pressure of low-carbon development on emerging economies and least developed countries is even higher. Regarding low-carbon technology in the ecological civilization, developed countries and developing countries face some common challenges. All countries need to cooperate to eliminate the bottlenecks in low-carbon technology development.

The inertia of high material consumption in industrial civilization is another important barrier for the green transformation. Industrialization has enabled the production of mass material wealth at low cost. This has provided basis for residents of developed countries that have completed the industrialization process a high level of material consumption. However, this consumption pattern and level are unhealthy and unsustainable. Low price of food with high calories and high fat content encourages people to eat more than they should, and when this causes health problems, they take large amount of prescribed medicine. People use high emission private cars as substitute for walking and then go to gyms to exercise with equipment to take away excessive calories. High calorie foods, prescribed medicine, gasoline powered vehicles, air conditioners, and gym venues are provided by industrialization and could not have existed in agricultural civilization. However, these kinds of high material consumption use fossil energy, emit greenhouse gases, and damage the environment. One thing they contribute, though, is stimulating economic growth. This kind of consumption is almost fixed in developed countries and hard to change. This consumption inertia is also playing a role in the emerging economies during the process of industrialization, constantly raising the level of fossil energy consumption and greenhouse gas emissions. If this consumption continues to be imitated by less developed countries with more than half of the world's population, the material basis of ecological civilization will be destabilized. To stop this unhealthy and unsustainable consumption inertia, developed countries need to take the lead, but developing countries also need to change their ideas and search for a healthy living style in order to accelerate the green transformation process.

Moreover, poverty traps in less developed countries are a counterpart to the high material consumption in developed countries. The Millennium Development Goal formulated by the United Nations in 2000 aimed to eliminate starvation and absolute poverty in developing countries. Fifteen years later, except for emerging economies, the poverty population and condition of other less developed countries have not had essentially changed. In 2000, the poverty population was mainly concentrated in rural areas. Fifteen years later, with the fast increase of urbanization in developing countries, poverty also partially shifted from rural areas to cities. On one hand, cities can generate the effects of aggregation and economies of scale and can provide basic social services and support the poor more easily. However, on the other hand, the shortage of job opportunities and resources in cities also makes it harder for the poor in the urban area to break away from the poverty trap. In 2012, the United Nations Conference on Sustainable Development in Rio de Jairo launched the process to formulate Global Sustainable Development Goal (SDG) as a continuity of the Millennium Development Goal. The SDG aims to eliminate absolute poverty by 2030 and to realize sustainable development. After the global financial crisis in 2008, economic growth and green transformation have slowed down in developed countries; emerging economies are confronted with the pressure of passing the middle-income trap; and less developed countries face huge material production and consumption barriers to eliminate poverty trap.

One of the technical barriers to ecological civilization transformation is the indicators and metrics of ecological prosperity. Agricultural civilization was a self-sufficient society, commodity economy was not advanced, and people traded goods for goods. Industrial civilization uses currency for measurement and developed a mature commodity economy. The System of National Accounts uses gross domestic product (GDP) or value added as measurements and adopts a single indicator with easy growth accounting methodology. For a long time, some people have been skeptical, or even critical, about the System of National Accounts, which honors the gold standard, especially from a sustainable development and ecology protection view. Yet people are still trying to measure ecological damage and ecological capital in monetary terms. This might lead to the misreading or misjudging of the value of ecosystem. Then, should the ecological prosperity of the ecological civilization era be measured with currency, or material object, or with multiple measurements and indicators? China advanced the notion to formulate balance sheets on natural resource assets. However, balance sheets are accounting statements of companies' business activities and are measured by market prices. Obviously simply applying the financial balance sheet is not ideal. Therefore, a breakthrough in the measurement method of ecological prosperity is of fundamental significance to the development of and transformation toward ecological civilization.

Governance problems and dilemmas are a systematic barrier to the transformation toward ecological civilization. There is no global government. This makes it hard to impose the systems, mechanism, rules, and regulations of the ecological civilization to sovereign nations and to put the ideas about ecological civilization into practice. In addition, the systems and power of voices of global governance are concentrated in a few developed countries who collude to block the intention of other countries for safeguarding their own interests. The practical needs for global sustainable development are then ignored.

The United Nations is a great platform for different voices. However, not all voices enjoy the same attention. The United Nation Millennium Development Goal is a global political consensus, but it only looks great on the outside with little implementation mechanisms. Negotiations on global climate change and the formulation of the global Sustainable Development Goal are all participative. The practice of green transformation still lacks a unified standard and concrete measures. At the state level, although there is support by the government, ecological civilization transformation is still faced with challenges. Governments follow a tenure system and tend to focus on short-term actions. Governments also have multiple departments and regions, which lead to conflicts in interests that are hard to harmonize. In addition, different stakeholders may have different demands and requests for the ecological civilization systems and mechanisms. Consumer behavior change is also impeded by constraints from existing social systems and mechanisms.

10.4 Early Practices

Ecological civilization, as a social civilization form, is the destined direction for social development. Compared with the transformation from agricultural to industrial civilization, the transformation from industrial to ecological civilization is faced with greater challenges and needs longer time commitment. China proactively seeks the transformation to ecological civilization; this has great significance for the green development and transformation of developed countries and other developing countries. As mankind enters the era of ecological prosperity with a steady-state economy, the transformation of many social factors is a natural process. The process, however, can be accelerated by social choices and policy guidance. Agricultural civilization could only be passively controlled by the rule of the Malthusian Principle of Population. However, industrial civilization dispelled the myth of population explosion. During the early stage of industrialization, countries didn't intervene with population growth. In the later stages of industrialization, the total fertility rate of developed countries has declined. Low birth rate and low death rate naturally stop population from further expansion, and this leads to the advent of preferential policies to encourage people having more babies, which has not achieved obvious results. This process has lasted for several centuries in Europe and around 60 years in Japan.

China launched and accelerated industrialization under the background of deficiency in capital and technology. In the early 1980s, China launched a powerful population intervention policy, which has slowed down population growth in China and dramatically changed China's population profile in as short as three decades. This enabled the accumulation of material wealth and boosted consumption level increase, thus laying a solid foundation for the transformation to ecological civilization. The 30 years of family planning policy implementation have slowed down the population growth rate, lowered the overall scale of population expansion, and made the Malthusian theory lose its power. Without the family planning policy, China's population would have reached 1.6–1.8 billion, 0.3–0.5 billion higher than it is today. The smaller actual population increase has not only eased pressure on resources and the environment but also played a great role in promoting social civilization transformation.

Other developing countries still have a high population growth rate. If the population is left to change naturally, like the process of developed countries, it may take a long time, the process will be slow, and the pressure on resources will be high, involving high uncertainties. China's family planning policy is proactive, responsible, and with significant foresights, but during its initial implementation, the policy was confronted with much skepticism, criticism, resistance, and obstacles. Now China has loosened up its one-child policy and allows couples meeting certain criteria to have two children. People's desire for children and family size expectation has changed, and the population profile and changing trends is similar to the developed countries in postindustrialization stages. In terms of urbanization process, developed countries have gone through urbanization and counter-urbanization. This process enables better social service, more effective usage of resources, and is overall more conducive to the introduction of ecological civilization. China has accelerated urbanization since its reform and opening up, and it will not follow suite with the developed countries and experience counter-urbanization in the later stages of industrialization. China will steadily advance urbanization and make the most of the scale and aggregation effect of urban areas, promote resource efficiency, reduce the pressure of environmental damage, as well as protect and increase ecological assets.

All cultures encourage innovation, but there are different options in the technology of innovation. Mankind faces two options in technology path selection. The first option is regular technology, which is to increase efficiency, reduce consumption, lower cost, and utilize marginal resources. The second option is revolutionary technology, which includes technologies for sustainable substitutes of fossil fuel and eliminates the dependency on unsustainable resources. The path taken by industrialized countries on technological innovation focuses more on the research and utilization of regular technology. This is also due to the industrial civilization's value choice of utility maximization. In energy technology, developed countries have invested many efforts in regular technology research and development. Methods have been tried out to increase energy efficiency and utilize marginal resources, attempting to double, quadruple, or even octuple the current energy efficiency.

In order to reduce the dependency on regular fossil energy import, the United States are developing its shale gas resources, which will bring in the shale gas revolution in the field of fossil energy. However, this kind of technology choice has "rebound effects," which will to some extent offset the resource-saving effects of the increase in technical efficiency. In addition, marginal resources are limited. Shale gas is also nonrenewable. Even if its exploitation will not impose environmental threats, the resource will be exhausted. Developed countries are also researching and developing revolutionary technologies, which include wind energy, solar energy, and biomass energy. However, under the market economy condition embedded with the industrial civilization concept of "profit maximization," these revolutionary technologies are slow to progress. If the technological development of developed countries mainly relies on technological upgrades, then that of China mainly adopts both technological upgrades and technological revolutions. On one hand, China relies on fossil fuel and tries to improve its energy efficiency. In the course of industrialization, China has made rapid progress in energy efficiency improvement and leaped to be among the world top ranks in vehicle fuel efficiency, thermal power conversion, building energy, and manufacturing sector energy efficiency, such as iron and steel industry and cement industry. Meanwhile China is investigating oil and gas reserves in seabed and developing and utilizing low-quality coal, to increase the use of marginal energy resources. On the other hand, China is putting more efforts in the development and utilization of renewable energies. China's exploring hydroelectric and rural biogas resources before 2000 aimed to address fossil energy resource shortage. Then in the twenty-first century, China's large-scale investment in and utilization of renewable energy, such as wind, solar, and biomass, is a proactive effort for revolutionary energy transformation.

Since the 2010s, China's wind and photovoltaic power have quickly exceeded the level and scale of developed countries, leading the world energy production and consumption revolution. China's energy technology transformation has turned from regular technology upgrades to revolutionary technological breakthroughs. Fossil energy is the power base for industrial civilization, but an ecological civilization must rely on renewable energy resources. China's technological innovation of energy is a strong cornerstone for the construction of an ecological civilization. For other developing countries, the old ways of fossil energy under industrial civilization will lead to a dead end plagued by practical problems such as resource exhaustion, environmental contamination, and difficulties in reducing carbon dioxide emissions. Instead of trying a road that leads to nowhere, it is better to refer to the practice and experience of China's ecological civilization creation and to develop and utilize renewable energy resources and technologies.

The change in social civilization form needs support from newer systems, mechanisms, rules, and regulations. Developed countries have experienced many shortcomings of the industrial civilization during the industrialization process and tried many systemic innovations, such as establishing environmental protection institutions and issuing related rules and regulations. These systems are also being applied in China and many other developing countries and have had some effect. However, they are not enough to guarantee the need for sustainable development and accelerate the transformation to ecological civilization. Therefore, China needs further systemic innovations.

In the first few years of the new millennium, China formulated the strategy of promoting the construction of ecological civilization. This has played a positive role in deepening the introspection on industrial civilization and exploring the new ways of ecological civilization. However, little progress has been made in systemic and mechanistic innovation. In the 2010s, Chinese government systematically formulated its vision for the institutional innovation of ecological civilization. This includes implementing the most rigorous resource protection system, damage compensation system, responsibility investigation system, as well as environmental management and ecological restoration system. In this way, the ecological environment is protected by systems of government. Other such systems include improving the property right system of natural resources and usage regulation system, designating ecological protection red line, exploring and formulating balance sheets on natural resource assets, and conducting departure auditing for leaders on natural resource assets. To establish lifelong responsibility system for ecological and environmental damage, implement paid usage for resources and ecological compensation, and reform the management system on ecology and environmental protection are also included.

Some of these new ideas utilized and deepened the system and mechanisms of industrial civilization, such as the property right system of natural resources and usage regulation system; some are new arrangements for ecological civilization, such as the ecological red line and the eco-compensation systems. Others need further exploration before actual implementation, such as formulating balance sheets on natural resource assets. Systems and mechanisms of the industrial civilization were established on those of the agricultural civilization and took a long period of time. Therefore, we shouldn't expect that those of the ecological civilization could be accomplished at one stroke. The essence lies in exploration, practice, and change in routine thoughts. When capitalist countries were confronted with periodic economic crisis caused by their economies approaching saturation, they sought to benefit themselves at other countries' expense. Their own benefits were maximized by quantitative easing of monetary policy and trade barrier policies. In the 2008 global financial crisis, China also adopted the policy of increasing public investment, a policy often used by developed countries in time of economy recession. Facts have proved that this policy is not totally conducive to ecological civilization transformation. During the 13th Five-Year Plan (2016–2020) of China's national economy and social development, rapid economic growth can still be expected. However, the construction of ecological civilization, the growth of population and economy, the ecological redline threshold in ecological civilization system innovations, and the implementation of eco-compensation indicate that China's economy is entering a new status: its population is nearing zero growth, its labor supply is becoming less than demand, the population is aging, the economic growth rate is slowing down sharply, social material wealth remains at a relatively high level, and the ecological environment is undergoing overall improvement. In other words, China will not adopt developed countries' practices of lifting their economy out of recession at the expense of other countries and shifting economic crisis. Instead, China agrees with and will control the new status of its economy and promote the civilization transformation.

[illegible]

[illegible] the ecological [illegible] of ecological civilization system innovation, and the implementation of [illegible] indicate that China's population is [illegible] population is near zero growth [illegible] labor supply [illegible] the population is aging, the economic growth rate is slowing [illegible] material [illegible] at a high level, and the ecological environment [illegible]. In other words, China [illegible] developed countries [illegible] economy [illegible] of other countries [illegible] crisis [illegible] control the new [illegible] and [illegible] the civilization [illegible]

Epilogue

China's transition toward ecological civilization is a huge systematic project that covers various social and economic aspects. It is an important academic, strategic, and practical subject that needs in-depth theoretic and practical research. As a researcher in a Chinese national research institute on social and human sciences, the author's work has received support from many Chinese and international organizations, including the China National Science Foundation, China National Social Science Foundation, the Science and Technology Support Program, the 973 Program, special funding from the Chinese public finance, the Philosophy and Society Innovation Project of the Chinese Academy of Social Sciences, the Chinese National Development and Reform Commission, the Chinese Ministry of Environmental Protection, the Chinese Ministry of Industry and Information Technology, the Chinese Meteorology Administration as well as local governments, the UK Overseas Development Fund, the Swiss International Development Agency, the US Energy Foundation, Foundation of the United Nations Intergovernmental Panel on Climate Change, as well as the China Energy Foundation Commission of Hong Kong, China. They have funded the research, field studies, and academic exchanges related to this book.

More importantly, the colleagues and students in my research team, including Hongbo Chen, Ying Chen, Shouxian Zhu, Guiyang Zhuang, Qing Li, Mou Wang, Yan Zheng, Meng Li, Zhe Liu, Changsong Liu, Changyi Liu, XinluXie, Ran Wang et al., have participated in the research, shared their academic ideas, written research reports, processed data and information, and made great contributions to the preparation of this book. They are not ordinary members of my research team but have written and published relevant research reports and theses as coauthors or independent authors. The contents about climate capacity, ecological civilization definition, industrialization progress, and the Yanjiao Case Study, as well as the contents about ecological system innovation, are partially based on the materials or fruits of my joint research with them.

Apart from the research team, the preparation of this book has also benefited from an efficient and responsible support team, including Li He, Peiran Peng,

J. Pan, *China's Environmental Governing and Ecological Civilization*,
China Insights, DOI 10.1007/978-3-662-47429-7

WeijingHao, Jianbin Xiong, and Supeng Xue. Especially my mother, who is already over 80 years old, offered great understanding and support to my work. My brothers and sisters looked after our mother and tolerated my lack of time to visit and look after her. Additionally, my wife, Yaping Du, took care of all household work and daily chores so that I could concentrate on working on this book, and my daughter also gave great understanding and support to my work.

Here I would like to express my sincere appreciation to all the Chinese and foreign agencies and organizations, senior and young members of my research team, the support team members, as well as my family for their great support in the preparation of this book.

Index

A
农业文明; Agricultural civilization, 5, 29, 40–42, 44, 75, 129, 156, 165, 173, 177, 209, 212, 213, 215–217, 220
阿玛提亚· 森 ; Amartya Sen, 174

B
均衡配置; Balanced distribution, 122
自然资源资产负债表; Balance sheet for natural asset, 217, 220
美丽中国; Beautiful China, 21, 23, 25–27, 207
京津冀; Beijing-Tianjin-Hebei, 78, 104, 105, 183, 187
罗素; Bertrand Arthur William Russell, 31
沼气; Biogas, 72, 132, 139, 142, 143, 145, 146, 219

C
二氧化碳排放; Carbon dioxide emission, 22, 23, 136, 181, 219
承载能力; Carrying capacity, 1–3, 8–12, 14, 15, 18, 20–27, 36, 39, 48, 75, 79, 84, 113, 114, 160–162, 171, 172, 177, 189, 194, 210
达尔文; Charles Robert Darwin, 44
化学需氧量; Chemical oxygen demand (COD), 9, 68, 142, 182
循环经济; Circular economy, 71, 81, 87, 181
文明转型; Civilization transformation, 75, 216–218220, 221
气候容量; Climate capacity, 1, 7–18, 20, 25, 26, 79, 82, 99, 111
气候移民; Climate migrants, 1, 17
科布 - 道格拉斯生产函数; Cobb Douglas Production Function, 214
COD *See* 化学需氧量; Chemical oxygen demand (COD)
睡城; Commuter town, 98
比较利益; Comparative advantage, 111, 198
顺应自然; Complying nature, 35, 37, 192
生态退耕; Converting farmland for ecological use, 118
协同发展; Coordinated development, 19, 60, 61, 104

D
休谟; David Hume, 43
衍生容量; Derived capacity, 9, 13
发展边界; Development boundary, 209
发展范式; Development paradigm, 29, 33, 34, 38, 39, 42–45, 51, 55, 75, 76, 78, 81, 114, 129, 132, 133, 135, 136, 141, 147, 156
二元结构; Dual structure, 163

E
生态资产; Ecological asset, 161, 171, 192, 195, 218
生态容量; Ecological capacity, 1, 5, 8, 12, 15, 66, 79, 107, 189

J. Pan, *China's Environmental Governing and Ecological Civilization*, China Insights, DOI 10.1007/978-3-662-47429-7

生态文明(界定,内涵,定位); Ecological civilization, 1, 6, 7, 21–26, 37, 41, 43, 45, 47, 48, 55, 57, 60, 61, 81–83, 91, 96–100, 102, 107, 123, 124, 129, 135, 136, 138, 139, 143, 145–147, 153, 154, 160, 161, 165, 166, 169, 172, 176, 181, 182, 203, 205–220
生态补偿; Ecological compensation, 19, 45, 48, 49, 102, 178, 184, 192–196, 199, 206, 208, 220
生态危机; Ecological crisis, 32, 197
生态伦理; Ecological ethics, 32, 49
生态足迹; Ecological footprint, 2, 12, 21–23, 26
生态治理; Ecological governance, 202
生态增长; Ecological growth, 156–161, 163
生态公正; Ecological justice, 37, 38, 40, 44, 48, 49, 169–172, 178, 208
生态马克思主义; Ecological Marxism, 32
生态移民; Ecological migrant, 1, 19
生态繁荣; Ecological prosperity, 209–212, 216, 217
生态红线; Ecological redline threshold, 99, 102, 124, 185, 221
生态安全; Ecological security, 7, 21, 24, 27, 110, 124–127, 188–192, 206, 207, 209
生态社会主义; Ecological socialism, 32, 45
市民化; Eliminating hukou based discrimination in cities, 95
排污权交易; Emission trading, 194, 202, 205, 208
能源革命; Energy revolution, 135–141
能源安全; Energy security, 49, 55, 56, 76, 77, 86, 108, 110, 112, 120, 121, 125, 134, 138, 145
能源转型; Energy transformation, 138, 142, 144, 219
环境容量; Environmental capacity, 2, 7, 9, 11–13, 17, 21, 25, 56, 67, 79, 83, 84, 99, 162, 176, 185, 187, 190, 191
外部性; Externality, 19, 121, 176

F
计划生育; Family planning, 213, 218
农耕文明; Farming civilization, 34, 35, 37, 181

G
政府购买; Governmental purchase, 206
绿色经济; Green economy, 40, 41, 109, 207, 208
绿色转型; Green transformation, 195, 209, 215–217

H
天人合一; Harmony between human and nature, 29, 32, 35–39, 43, 44, 47, 80, 83, 97, 140, 147, 151, 171, 177, 209
钱纳里; Hollis B. Chenery, 52
胡焕庸线; Hu Line, 3, 5–7, 83
户籍; Hukou, 76, 88–90, 94, 95, 104, 194
以人为本; Human orientation, 80, 170

I
工业文明; Industrial civilization, 14, 20, 21, 25, 42, 43, 45, 47, 51, 75, 76, 96, 98, 100, 123, 129, 133, 135, 139, 141, 147, 155–158, 160, 161, 165, 166, 169, 172–174, 176, 177, 181, 192, 200, 202–204, 209–217, 219, 220
制度创新; Institutional innovation 主体, 181–187, 203, 220
工业化; Industrialisation, 23, 29, 51–73, 77, 113, 129, 158, 172, 188, 213
工业革命; Industrial revolution, 29, 30, 40, 42, 43, 51, 54, 85, 129, 135–139, 144, 154, 215

J
边沁; Jeremy Bentham, 43
季羡林; Ji Xianlin, 35

K
凯恩斯主义; Keynesianism, 212

L
老子; Lao Tze, 35
期望寿命; Life expectancy, 31, 151, 167, 168, 174, 213
增长的极限; Limit of Growth, 34
区位优势; Location advantage, 97, 197, 199
低碳经济; Low-carbon economy, 40, 41, 81

M
功能区; Major functional zones, 114
马尔萨斯陷阱; Malthusian Trap, 156, 160, 162

物欲消费; Material desire and consumption, 168, 172–174, 176, 177
千年发展目标; Millennium Development Goal (MDG), 46, 108, 109, 216, 217

N
自然约束; Natural constraint, 152, 153, 169
负效用; Negative utility, 174–178
反效用; Neutralizing utility, 177
非化石能源; Non-fossil energy, 131, 146

O
物种起源; On the Origin of Species, 44

P
生态服务付费; Payment for ecological service, 19, 193
避风港; Pollution haven, 61
人口爆炸; Population explosion, 153, 156, 217
正效用; Positive utility, 174–176, 178
贫困陷阱; Poverty trap, 20, 215, 216
剪刀差; Price scissor, 119
资源消费的累进税; Progressive tax on resource consumption, 179
公众参与; Public participation, 83, 87, 116, 193, 206

Q
叶谦吉; Qianji Ye, 35
量化宽松; Quantitative easing, 157, 158, 220

R
蕾切尔· 卡逊; Rachel Carson, 30
理性消费; Rational consumption, 45, 49, 165, 169, 171–177
真实增长; Real growth, 160
反弹效应; Rebounding effect, 140, 153, 219
耕地红线; Redline minimum area threshold for arable land, 99
可再生能源; Renewable energy, 26, 44, 48, 49, 81, 86, 87, 129–132, 134, 136–139, 142–146, 172, 194, 205, 219
联; Resource nexus, 107–127
尊重自然; Respecting nature, 21, 24–26, 35, 37, 161, 165, 166
罗马俱乐部; Rome Club, 1

S
饱和; Saturation, 91, 151, 155, 156, 159, 160, 169, 174, 213, 214, 220
科学发展; Scientific development, 40, 79, 182
科学规划; Scientific planning, 82, 96, 98
寂静的春天; Silent Spring, 30, 34
雾霾; Smog, 36, 69, 72, 131, 134, 152, 162, 175, 183, 185, 187, 189–191
社会达尔文主义; Social Darwinism, 44
宙飞船经济; Spaceship economy 预, 3, 31
静态经济; Static economy, 30
恒态经济; Steady-state economy 污染, 31, 33, 147, 161, 164, 209, 212, 217
农业转移人口; Surplus population transfer from agriculture, 118
可持续消费; Sustainable consumption, 165, 177
可持续发展目标; Sustainable development goals (SDG), 109, 137, 216, 217

T
道德经; Tao TeChing, 35
技术进步; Technology progress, 2, 10, 12, 96, 135, 140, 151, 152, 166, 189
展; Transformation development 资源关, 75, 76, 78, 81
三重约束; Triple constraints, 151–156, 160, 161

U
城市病; Urban disease, 78, 79, 85, 87, 96, 98, 189
城镇化; Urbanization, 3, 16, 23, 24, 27, 29, 47, 55, 56, 58–63, 69, 75–83, 88, 89, 91, 92, 94–100, 102, 104, 113–115, 123, 129, 130, 132, 134, 135, 141, 144, 146, 156, 158, 159, 188, 191, 193, 195, 197, 204, 216, 218
功利主义; Utilitarian, 42–45

V
无知之幕; Veil of ignorance, 92, 170, 171

W
水安全; Water security, 12, 108, 110, 120–123, 125
世界工厂; World factory, 98

Y
长三角; Yangtze Delta, 3, 6, 47, 66, 78, 111
燕郊镇; Yanjiao Town, 102–105
易经; Yi-Ching, the book of change 宇, 33

Z
零和博弈; Zero-sum game, 25, 210